The Construction of the Heavens

The astronomical observations of William Herschel (1738–1822) made him question the accepted model of the clockwork universe. This volume explains the development of Herschel's thoughts on what he called 'the construction of the heavens' and reprints his principal papers on this subject.

The preliminary chapters provide an introduction to Herschel, including his unusual path to astronomy, the discovery of Uranus and his work on the evolution of stellar clusters, which eventually led him to challenge the unchanging Newtonian universe. The second half of the text comprises eight of Herschel's key papers on what we today would call cosmology, representing his progress between 1783 and 1814, fully annotated with historical notes and modern astrophysical explanations.

Ideal for undergraduate and postgraduate students in the history of science and in astronomy, this volume explains Herschel's pivotal role in the transformation from the clockwork universe to the 'biological' universe of modern astronomy.

The author of four previous books on the Herschels, Michael Hoskin has taught history of astronomy at the University of Cambridge throughout his career. He is an Honorary Fellow of the Royal Astronomical Society, Doggett Prizewinner of the American Astronomical Society, and Jaschek Prizewinner of the European Society for Astronomy in Culture. In acknowledgement of his contributions to the history of astronomy, the International Astronomical Union has named a minor planet Hoskin.

Cover illustration (front, inset): miniature of William Herschel as a young man, artist unknown, Herschel Family Archives; (front, main): William Herschel's diagram to illustrate his theory of the solar apex, from his paper in *Philosophical Transactions*, vol. 73 (1783); (back): watercolour that accompanied the 25-ft reflector that William Herschel made for the King of Spain, courtesy of the Observatorio Astronómico Nacional, Madrid.

The Construction of the Heavens

William Herschel's Cosmology

MICHAEL HOSKIN
Churchill College, Cambridge

with astrophysical notes by

DAVID DEWHIRST

AND

WOLFGANG STEINICKE

CAMBRIDGE
UNIVERSITY PRESS

CAMBRIDGE UNIVERSITY PRESS
Cambridge, New York, Melbourne, Madrid, Cape Town,
Singapore, São Paulo, Delhi, Tokyo, Mexico City

Cambridge University Press
The Edinburgh Building, Cambridge CB2 8RU, UK

Published in the United States of America by Cambridge University Press, New York

www.cambridge.org
Information on this title: www.cambridge.org/9781107018389

First published by Cambridge University Press 2012

Printed in the United Kingdom at the University Press, Cambridge

A catalogue record for this publication is available from the British Library

Library of Congress Cataloguing in Publication data
Hoskin, Michael A.
The construction of the heavens : William Herschel's Cosmology / Michael Hoskin ; with astrophysical notes by David Dewirst, Wolfgang Steinicke.
p. cm.
Includes bibliographical references and index.
ISBN 978-1-107-01838-9 (hardback)
1. Herschel, William, Sir, 1738–1822. 2. Astronomers – Great Britain – Biography. 3. Astronomy – History. 4. Cosmology – History. I. Dewirst, David. II. Steinicke, Wolfgang. III. Title.
QB36.H6H719 2012
520.92–dc23
[B]

2011036932

ISBN 978-1-107-01838-9 Hardback

Contents

Preface

Half-a-century ago, as a young historian of astronomy at Cambridge University, I published my first book, *William Herschel and the Construction of the Heavens*. Its theme was the investigations into the universe beyond the solar system that were carried out in the late eighteenth and early nineteenth centuries by the Hanoverian-born musician-turned-astronomer, William Herschel. Herschel marks the watershed separating the clockwork universe of Newton and Leibniz from the cosmos in which we live today, in which everything – even the cosmos itself – has a life history. Because Herschel's own writings on cosmology were difficult to access, being available only in the original pages of *Philosophical Transactions* or in the two massive volumes of collected works published in 1912, I also reproduced the most important of his publications on the subject. In this I was fortunate to enlist the collaboration of a Cambridge astrophysicist, David Dewhirst, who supplied extensive footnotes in which he identified the objects that Herschel was citing in his papers.

In the succeeding fifty years I have published four books and innumerable articles on William Herschel, his sister Caroline who was his collaborator, his son John, and the other members of this hugely-talented family. I have learned much in the process, and Part I of this book presents my current understanding of Herschel's work on what he termed "the construction of the heavens". Part II contains a slightly different selection of Herschel's own papers. Gone are two papers published when he was nearly eighty, which I now realize proved to be of minor significance; included are the remarks prefaced to his third and final catalogue of nebulae. After half-a-century, I once again have the collaboration of David Dewhirst, but this time he is joined by Wolfgang Steinicke, whose familiarity with the nebulae studied by Herschel and his immediate successors is unrivalled.

John Herschel-Shorland has been generous and helpful as always. I am grateful to the Royal Astronomical Society for permission to reproduce documents from their Herschel Archive, and to their Librarian, Peter Hingley, and to the Librarian of the Cambridge University Institute of Astronomy, Mark Hurn, for many favours. Bernard Hoskin kindly drew the diagrams I use to illustrate my argument. Finally, my thanks to my editors at Cambridge University Press, with whom it has been a pleasure to collaborate.

Introduction

William Herschel (1738–1822) was a musician and composer for the first half of his life, and astronomer to the King of Britain for the second half. Astronomers of the time might distinguish themselves either as makers of telescopes, or as observers, or as theoreticians. Herschel distinguished himself in all three.

In November 1778, while a musician in the English spa resort of Bath, Herschel as an amateur observer ground and polished for his 7-ft reflector a mirror that was simply the finest anywhere; and using it he discovered the planet we know as Uranus. This won him the patronage of the King and with it the opportunity to give up music and dedicate himself to astronomy. With funding from the King he then built himself the biggest reflector ever seen, and he conducted a brisk trade in telescopes, the crowned heads of Europe competing to be allowed to buy a Herschel reflector.

As an observer, Herschel saw himself as a natural historian of the heavens, collecting and classifying specimens in vast numbers. While still an amateur he began collecting double stars by the hundred, and as a professional he published catalogues of 2,500 nebulae and clusters where only a hundred or so had been known before.

But it is as a theoretician that Herschel earned his pre-eminent place in the history of astronomy. Isaac Newton had envisaged the universe as a hugely-complex piece of clockwork, the creation of God the Clockmaker. Within this universe nothing really changed, just as in a clock the hands move but nothing really changes. But there was a problem. Forces generate motions, and Newton claimed that gravity was a universal force; yet when Newton was in his prime it seemed that the stars were as ‘fixed’ and motionless as they had ever been. His way out of the difficulty – known only to his intimates – was to argue that the stars were distributed (more or less) uniformly throughout infinite space, so that each star was pulled (more or less) equally in every direction by the other stars and thus remained at rest, at least in the short term. Over the long term (he believed) things did not go so well, and so from time to time Providence had to intervene and restore the original order to the universe of stars.

In 1785 Herschel invited the readers of *Philosophical Transactions* to imagine a universe that likewise began with a near-uniform distribution of stars throughout an indefinite space. Although Newton had never given a demonstration that gravitational attraction operated beyond the solar system, Herschel found clear proof of this in the existence of star clusters: the clusters must surely have formed as previously-isolated stars attracted each other and moved ever closer together.

Significantly, in some of the clusters that he encountered the stars were scattered, while in others they were tightly packed; and surely *in time* the scattered clusters would become ever more compact, as gravity continued to pull the stars towards each other. In other words, scattered clusters were young, and compact clusters were old. This insight, that individual clusters have a life history – that cosmogony is the key to the study of what Herschel termed "the construction of the heavens" – tolled the death-knell of the clockwork universe.

In his early years as a professional astronomer Herschel envisaged our Galaxy as a layer or 'stratum' of stars of limited extent, and, as we shall see, he devised a novel method for plotting its outline. There were also, he thought, other strata in the sky that were so extensive that they might even "outvie our milky-way in grandeur": these were what we would term 'galaxies'. Later he was forced to admit that the stratum of our Galaxy seemed in fact to be without bounds, and if so it was unique. Not only that, but it was evidently starting to fragment under the destructive force of gravity, and so its future existence was limited in time – and so, intriguingly, must be its past.

In the later 1780s Herschel convinced himself that the luminous bodies in the larger universe are exclusively stars, isolated or in clusters, but in 1790 a new observation persuaded him of the existence of 'true nebulosity', a luminous fluid out of which stars were born. This led him to take his cosmogony back in time to the period before stars had formed; in imagination he allowed gravity to work its magic first on this fluid, and then on the resulting stars. As it was not possible in the brief span of human life for an individual person to observe this evolution unfolding, in papers published in 1811 and 1814 Herschel offered the equivalent by arranging his nebulae and clusters sequentially by age, into 'articles'. These, he said, were comparable to a succession of portraits of an individual man as he went through life.

Like all astronomers of the period, Herschel knew that the speed of light though great was finite, and he was the first clearly to state what has become a commonplace in modern cosmology, that to see an object at distance is to see it in past time. Indeed, he believed that he personally had seen light that had been two million years on its way, and that if the source of the light had been annihilated soon after the light departed on its journey, Herschel would still see the object when its light eventually reached him.

In his years as an amateur astronomer Herschel had focused his efforts on the nearer stars. As he examined them one by one, he listed those that were 'double', two stars so close that at first inspection they appeared to be one – for he knew that Galileo had popularized a method of using doubles to establish the distances of the nearer of the two stars. But when he re-examined some of his doubles after another twenty or so years, he had a quite different reward for his labours: in some of the pairs, the two stars had performed a gyration around each other – they were companions in space, bound together by an attractive force, no doubt gravity. This was the first time that such a force had actually been seen in action in the stellar universe.

In these investigations, whether of double stars or of nebulae and clusters, Herschel's telescopes gave him the advantage over all other astronomers, for much of the evidence was available to him alone. But there was one question where he relied exclusively on evidence available to all: the direction in which the solar system is travelling through space. By the 1780s a small number of stars were known to be in motion, moving across the sky almost imperceptibly year by year; and surely the star we call the Sun (accompanied by its planets) must also be in motion. Now if we inhabitants of the solar system are travelling in a certain direction, we will expect the stars that we are approaching to appear to move to one side or the other, as do the trees that are in front of us as we walk towards a forest. Herschel was the first to identify such a pattern and to propose the actual direction in which the solar system is moving.

Astronomers of Herschel's day, amateur and professional alike, were preoccupied with the solar system and believed themselves to inhabit a clockwork universe. Today's universe could not be more different: the solar system is a minority interest, and we see ourselves as living in a 'biological' cosmos that itself was born out of a Big Bang. Herschel's published articles in *Philosophical Transactions* began this seismic shift. This book explains how his thinking developed, and reprints these articles, with footnotes drawing on what we have since learned about the objects he discusses.

PART I HERSCHEL'S EXPLORATION OF THE COSMOS

1

The making of an astronomer

William Herschel (Fig. 1.1) was born in Hanover on 15 November 1738 into a humble but musically talented family.[1] His father Isaac was a bandsman in the Hanoverian Guards,[2] and Isaac's infant son was baptized Friedrich Wilhelm Herschel in the Garrison Chapel. We know him as William because this is the name he always used after he fled to England in 1757 to escape the aftermath of the Battle of Hastenbeck, and in 1793 the name became legally his by Act of Parliament.[3]

All the Herschel boys grew up to be musicians able to perform professionally on a range of instruments, and for more than half his life Herschel was a violinist, oboist, harpsichordist, organist and singer, whose ambition was to be remembered as the composer of symphonies and concertos. After his arrival in England, Herschel performed and taught music, first in London and then in the north of the country. In 1766 his fortunes took a turn for the better, when he was invited to become organist of the Octagon Chapel then under construction in the fashionable city of Bath. There the aristocracy came to take the waters and enjoy musical entertainment during the season, which lasted from the autumn through to Easter, and so there were rich pickings in Bath for an enterprising and hard-working musician – although the intense competition could sometimes lead to unseemly squabbles. Herschel soon established himself there as one of the two leading performers, and before long he was joined by his younger brother Alexander, a cellist whose remarkable talent as a brass-worker was later to prove invaluable in the construction of telescopes.

Herschel had two surviving sisters. The elder, Sophia, was long since married and had a large family of her own, but Caroline, eleven years William Herschel's junior, was languishing as a household drudge in the family home in Hanover. Their mother, Anna, enjoyed having reliable and unpaid help around the house

Fig. 1.1 Miniature of William Herschel as a young man, artist unknown. Herschel probably gave it to his sister Caroline when he travelled from England to visit the family home in Hanover in 1764. In her old age, Caroline gave it to her brother's only son John. Herschel Family Archives.

and she did everything in her power to prevent Caroline acquiring the skills – in needlework, or in the French language – that might have allowed her to make her escape.[4] Caroline was diminutive and marked with the scars of smallpox, and her father (who died in 1767 of health problems brought on by war service) had long ago warned her not to expect an offer of matrimony. To bring Caroline to Bath her brothers invented an implausible excuse – that she might prove to have a good singing voice, and if so could play a role in the Handel oratorios that Herschel promoted. Anna was won over by the promise of an annuity to pay for substitute help in the home, and so it was that in 1772 Caroline joined her brothers in Bath.

It turned out that she could indeed sing (Fig. 1.2). But she had limitations. In childhood, as a female pupil of the Garrison School in Hanover, she had not been allowed to learn arithmetic, and so could not keep the household accounts – let alone play a future role in the astronomy that was fast becoming

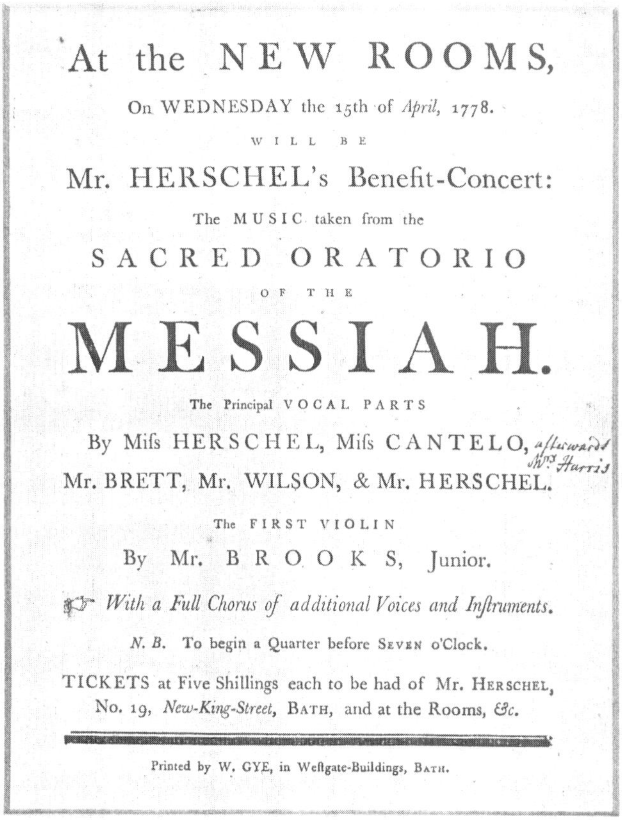

Fig. 1.2 Poster advertising a performance of Handel's *Messiah* in Bath, 15 April 1778. It was a 'benefit concert' promoted by Herschel, that is, one for which he took full financial responsibility, and both he and his sister Caroline performed as soloists. After the performance Caroline was invited to perform at Birmingham, an engagement that might have launched her independent career as a singer, but she declined. Herschel Family Archives.

a passion for her older brother. Herschel lost no time in remedying her lack of arithmetic. He also taught her some elementary geometry, the hard way: when cutting herself a slice of pudding, if she failed to estimate correctly the angle made by the slice, this grown woman was made to go hungry.

Although when Herschel was a boy his father had rarely had two *gutengroschen* to rub together, the admirable Isaac had somehow learned something about the ideas of Newton, Leibniz and Euler, and had shared these with his sons, not as truths to be learned but as subjects for critical debate. As a boy bandsman in the Hanoverian Guards Herschel had used his army pay to purchase the three volumes of John Locke's *On Human Understanding*. Later, when he was a musician in the north of England, Herschel had bought *Harmonics* by the Cambridge professor Robert Smith; and in Bath he decided to buy the same author's two-volume treatise on *Opticks*.[5] This told him a lot about how to build a telescope, and a little about what to see with the completed instrument. His appetite whetted, soon after Caroline's arrival he purchased the very successful introductory text *Astronomy* by the sometime shepherd-boy James Ferguson.[6] Although no more than a popularizer, Ferguson had been elected a Fellow of the Royal Society and granted a small 'pension' by the King, and Herschel therefore took him very seriously.

In the first edition of his book, Ferguson had little to say about the stars. Their Greek name of 'fixed stars' was still widely used, although everyone now knew that they were in fact isolated bodies in space and able to move freely. For most purposes they were still an unchanging and therefore uninteresting backcloth to the motions of the planets, moons and comets of the solar system; and so it was these nearer bodies that continued to preoccupy astronomers. Fortunately it was a later edition of *Astronomy* that Herschel had bought, one to which Ferguson had added a chapter "Of the fixed Stars".

Herschel was eager to see the sights described by Smith and Ferguson, and for this he needed telescopes.[7] He first experimented with refractors, in which a lens at the upper end of the tube bends the incoming light and brings it to a focus near the bottom where it is examined through an eyepiece. The problem was that the different colours are bent ('refracted') by the lens through slightly different angles and so come to a focus at slightly different distances. This results in a blurring ('chromatic aberration'), which can be reduced by making the tube as long as possible. But a telescope of excessive length is hard to direct and control, and Herschel soon abandoned refractors. The solution, as Newton had shown, was to have at the foot of the tube a mirror that reflected the light back to a focus near the top; this avoided the problem because all colours are reflected equally. At the top of the tube, a 'Newtonian' reflector has a small flat mirror set at an angle, and this directs the image sideways, where the observer examines it through the eyepiece. Somewhat surprisingly, therefore, in a Newtonian reflector the observer is positioned at the top of the tube looking sideways to the direction from which the light is coming. Guided by what Smith had to say, Herschel made himself reflectors of ever-increasing size: 5½-ft focal length, then 7-ft, then 10-ft, and finally, in July 1776, 20-ft. Mirrors that were of the same diameter and focal length were interchangeable, and Herschel experimented endlessly, grinding and polishing in search of the ideal shape and reflectivity, while Caroline read to him to relieve the boredom and sometimes put food into his mouth as though feeding a baby.

The larger the mirror, the more light it collected. This facilitated greater magnifications; more importantly for Herschel, it also brought into view faint objects invisible to the human eye in smaller instruments. But whereas he would soon devise practical wooden stands for his 7-ft and 10-ft reflectors, the 20-ft with its mirrors 12 inches in diameter was simply slung from a pole, with the observer on a ladder alongside.

In his years in Bath as an amateur observer, Herschel was a busy musician whose leisure hours were mostly given to improving his telescopes.[8] When he did make time to observe he normally used the 7-ft or the 10-ft rather than the unwieldy 20-ft. In March 1774 he had felt confident enough to open his first observing book, and like amateurs ever since the first objects he observed were Saturn and the Orion Nebula.[9] It happened that the ring of Saturn was then edge-on to us, and when this occurs moons that are normally lost in the glare of the ring may become visible. Saturn became an abiding interest of his, and fifteen years later, when the ring was again edge-on, Herschel would discover its sixth and seventh moons.

In his Bath years Herschel returned to the Orion Nebula from time to time to see if it had altered shape, and of this more anon. He studied the Moon in the hopes of finding evidence of the inhabitants whom Ferguson called the Lunarians, and when such evidence proved hard to find he turned to measurements of the heights of the lunar mountains.

Scientifically his isolation ended in December 1779 when he was invited by a Bath physician, Dr William Watson Jr, to become a member of the newly-founded Bath Philosophical Society.[10] There he met kindred spirits, and Herschel read to the society innumerable papers on all kinds of subject, beginning with – of all things – corallines. By now, word of this indefatigable amateur astronomer was beginning to spread. The Astronomer Royal, Nevil Maskelyne, had earlier called at Herschel's home, and the two men indulged in what Caroline described as "several hours spirited conversation", so spirited that Caroline feared they were quarrelling.[11] Watson and Maskelyne became Herschel's devoted allies, who recognized his unique gifts, made him known in Royal Society circles, and began – in partnership with Sir Joseph Banks, president of the Society – to plot ways by which Herschel might be given the opportunity to devote his life to astronomy. Meanwhile he was exploiting the possibilities opened up by the superb mirror he had made in November 1778 for his 7-ft: in August 1779 he began using it systematically to 'review' all the stars down to eighth magnitude (Ptolemy had classed the faintest naked-eye stars as sixth magnitude), looking for those that appear single at first glance but are in fact double.

The enormous distances of even our nearest neighbours among the stars had been estimated by Newton and others on the assumption that the fainter the star, the more distant it is likely to be. But this was no substitute for actual measurements, and it had long been recognized that these would be delicate almost beyond belief. Herschel knew that Galileo had popularized a method for simplifying the daunting technical problems: the astronomer should study pairs of stars that happened by chance to lie in the same direction from us. If one star of the 'double' (no doubt the fainter) was at a great distance, it would serve as a virtually-fixed reference point, against which the astronomer could measure the apparent movement of the nearer star as we orbit the Sun. This apparent movement would reveal the distance of the star: the greater the movement, the nearer the star.

Herschel hoped his double stars might enable other astronomers to pursue the problem of stellar distances. This did not work out as planned, but the review had other consequences, of the greatest importance. First, it demonstrated that Herschel was a dedicated observer of exceptional skill. The catalogue of double stars he sent to the Royal Society at the end of 1781 contained no fewer than 269 entries, many of them doubles that appeared single in the best telescopes that existed elsewhere in England. In particular, Herschel declared that the Pole Star was double, but it was weeks before any other observer could confirm this, and when confirmation was at last forthcoming Banks wrote to Herschel to offer his congratulations.[12]

Second, on 13 March 1781, Herschel's review brought him to a 'star' that he instantly recognized was anomalous, and so he returned to it four days later, to find that it had moved: it was therefore a member of the solar system, presumably a comet. In the best telescopes at Greenwich and Oxford it looked like any other star, but when it was at last identified and its orbit had been tracked and analysed by mathematicians it proved to be a planet orbiting the Sun beyond Saturn, the first planet to be discovered since the dawn of history.[13]

This was the opportunity Herschel's allies had been looking for. If King George III, who was not only King of Britain but Elector of Herschel's native Hanover, could be persuaded to allow Herschel to name the planet in his honour, the customs of patronage would require the King to show his appreciation in financial terms. The King was understandably attracted by the thought that, from now on, the planets of the solar system would be Mercury, Venus, Earth, Mars, Jupiter, Saturn, George; but he was at a loss as to how to respond. Eventually it was agreed that Herschel would move to the neighbourhood of Windsor Castle, and would become Astronomer to the King, with a 'pension' that would leave him free to dedicate his time to astronomy except when called upon by the Royal Family or their guests. And so, in August 1782, with his life more than half run, Herschel became a professional astronomer at last.

2

Our neighbours among the stars

The primary motivation for William Herschel in his career as a professional astronomer was his passion to understand what he termed "the construction of the heavens", the large-scale structure of the universe of stars. But the study of distant and faint objects called for reflectors with large mirrors that had great "light-gathering power", and it was not until late in 1783 that he built himself the instrument he needed, a 20-ft with 18-inch mirrors and a stable mounting. As an amateur, therefore, and as a newly-fledged professional, he explored a number of unresolved questions concerning our nearest neighbours in the starry universe. How far away are the nearest stars, and are there perhaps pairs of stars at the same distance, bound to each other by gravitational attraction? Why do some stars vary in brightness? In which direction is the solar system travelling through space? And which stars (if any) accompany the Sun and so make up the cluster of which the Sun is a member?

The distances of the stars

The distances of the stars was a problem that had engaged astronomers since Antiquity, and in the late sixteenth century it became a point of contention in the light of Copernicus's claim that the Earth was not the centre of the cosmos but was in fact in annual orbit around the Sun. An obvious objection was, "If we are (as you say) viewing the stars from a moving platform, why then do they not appear to us to move?" The response of the Copernicans could only be: "Unfortunately the stars are so very far away and their apparent movements so tiny that our present instruments are not sensitive enough to detect them."

At first this seemed no more than a wilful refusal to accept decisive observational proof of the stability of the Earth, especially when the Danish observer

Tycho Brahe brought the accuracy of his instruments to near the limit of what is possible with the naked eye. But in the seventeenth century the question was transformed, as developments in dynamical thinking convinced astronomers that the massive Sun was a more reasonable choice for the centre of the planetary system than the tiny Earth. The search for apparent annual movements ('annual parallax') among the stars was now driven by the desire of observers to determine the scale of the stellar universe, for the greater a star's apparent movement, the nearer to Earth it will be.

Isaac Newton revealed the extreme delicacy of the measurements involved when he made the working assumption that the Sun is physically similar to Sirius, the brightest star (other than the Sun itself). This being so, a determination of the ratio of their brightnesses would reveal the ratio of their distances from Earth. The apparent brightness of the Sun, he found, was some million million times that of Sirius, and since light falls off with the square of the distance, Sirius must be some one million times further from us than the Sun.[1] If so, the apparent movement of Sirius as seen from Earth as it orbits the Sun would be no more than the width of a coin several miles away.

Worse still, the observations required to detect such a tiny movement would be spread over many months, during which time seasonal changes in temperature and humidity might well impact on the instrument being used. Not only this, but variations in atmospheric refraction – whose effect was then imperfectly understood – would distort the measurements.

One way to circumvent the problem of atmospheric refraction would be to observe the 'target' star only when it was exactly overhead and its light was entering the Earth's atmosphere at right angles. Robert Hooke noticed that the star γ Draconis passed directly overhead his house in London, and in 1669 he half-heartedly made observations to try and detect its parallax (Fig. 2.1).[2] The idea was taken up in more committed fashion in the mid-1720s by the amateur observer Samuel Molyneux and his friend James Bradley, future Astronomer Royal, who both had telescopes specially designed and built and mounted rigidly against masonry walls. Molyneux died in 1728, but Bradley completed a campaign of meticulous observations and found, not the parallactic effect he was looking for, but the 'aberration of light', the alteration in the observed position of a star caused by the *velocity* of the Earth's movement through space as it orbits the Sun. Even when this was taken into account, Bradley's telescope was still not accurate enough to detect annual parallax. Knowing what its accuracy was, he estimated that γ Draconis must be at least 400,000 times the distance of the Sun[3] (or 32 million million miles, as Herschel learned from Ferguson's textbook[4]). Smith's *Opticks* declared the stars nearest to the solar system to be so far away that even the vast orbit of the Earth around the

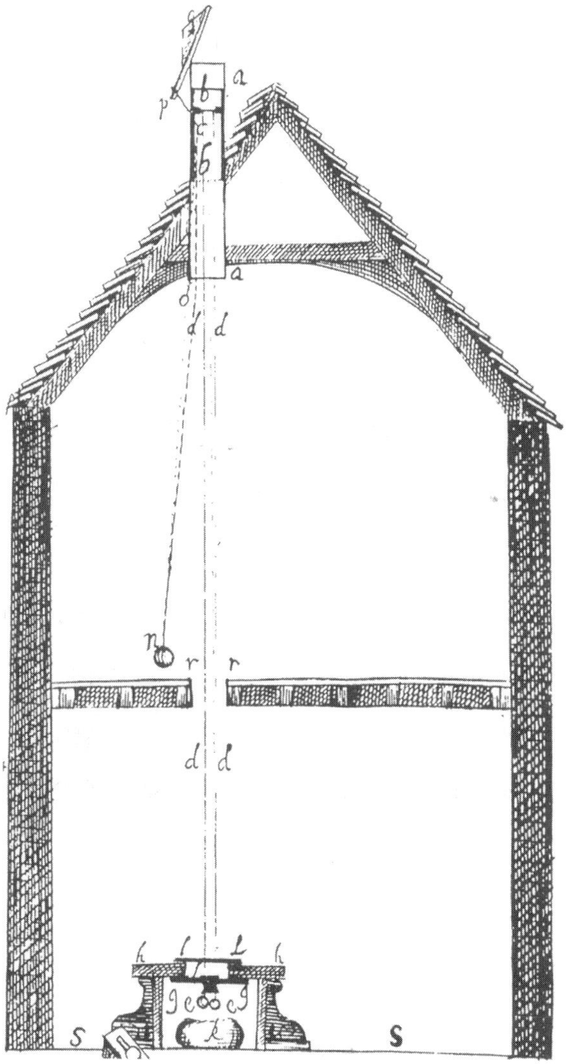

Fig. 2.1 Robert Hooke's zenith telescope, incorporated into the fabric of his London house to ensure its stability during observations that were planned to extend over many months. It is a remarkable example of a telescope built to observe just one single star (γ Draconis) and only when this star was directly overhead. From Hooke's *An Attempt to Prove the Motion of the Earth* (London, 1674).

Sun, as seen from a star, would be imperceptible, "such an amazing conclusion as could not be believed, were it not supported by undeniable evidence".[5]

There was another way to circumvent the daunting problems facing astronomers attempting to measure annual parallax, the method of double stars. This had been popularized in 1632 by Galileo (although it had in fact been proposed earlier by his little-known contemporary Lodovico Ramponi).[6] Suppose two stars lie in almost exactly the same direction from Earth, so forming a 'double star'; and suppose that one of them – no doubt the brighter – is much the nearer of the two. The stars, being so close in the sky as seen by us, will be almost equally affected by refraction and any similar complications, and by any warping of the instrument. But their apparent movements will be directly proportional to their distances, and therefore very different. Indeed, if the further star is sufficiently far away it will be for all practical purposes a fixed reference point placed in the heavens by a helpful Nature. This being so, the observer need only track changes in the tiny angle between the two stars.

In 1747 Bradley announced his discovery of yet another complication facing observers eager to detect annual parallax: the 'nutation' or wobbling of the Earth's axis caused by the gravitational attraction of the Moon on the Earth (which is not a perfect sphere). Double stars seemed to him to offer the most promising way forward, and so he urged astronomers to "examine nicely the relative Situations of particular Stars: and especially those of the greatest Lustre, which, it may be presumed lie near to us, and may therefore be subject to more sensible Changes", and at the same time to "determine the relative Positions of some of the *smallest* [and presumably very distant] that appear near them".[7]

Neither Smith nor Ferguson had anything to say about this method of double stars, and there is no reason to think that Herschel had read Bradley's paper by late January 1778, when we find him interrupting his records of observations to write out a mini-treatise concerning the method and its advantages.[8] It is a surprisingly competent exposition, and it is hard to believe that he dreamt it all up for himself. It is much more likely that he had learned the method from the Astronomer Royal, Nevil Maskelyne, when he called on Herschel in 1777 (or perhaps 1776).[9]

At all events, on 28 January 1778 Herschel examined Sirius "But could not discover any small star near enough for the purpose of Parallax", and on 7 February he did the same for Procyon, with the same result.[10] On 8 April he noticed that Castor was double, but he evidently thought the components too similar in brightness (and therefore likely to be at too similar a distance) to be of use. And there the matter rested for the time being.

Herschel then decided it was time for him to familiarize himself with all the brightest stars, and he looked in turn at each star of the first four magnitudes.[11] This 'review' was an elementary exercise and left no trace in his observing book. The second review, however, which he began on 17 August 1779, was a more serious matter. It extended to stars down to the eighth magnitude, and now he was on the hunt for double stars that might help the search for annual parallax. The review took him over two years and resulted in his first catalogue of 269 double stars, 227 of which had previously gone unrecognized. The paper was 'read' to the Royal Society in January 1782 and shortly thereafter published in *Philosophical Transactions*.[12]

Herschel had explained the motives behind his current investigations in an earlier paper "On the Parallax of the Fixed Stars", read to the Royal Society a few weeks before, in which he incorporated large sections of the mini-treatise he had written in the privacy of his observing book more than two years earlier.[13] On 19 April 1782, Maskelyne wrote him a long and friendly letter reporting the views of the Royal Society Publications Committee and the improvements they suggested in the paper before it would be considered ready for publication.[14] In particular they were unhappy about two postulates on which Herschel based his mathematical treatment. Under the first Herschel assumed that the stars were physically much the same, so that the brighter ones were always the nearer. Under the second he assumed that a star classified as of the n-th magnitude was n times further than a star of the first. The Committee considered the first postulate unnecessarily strong, and the second positively false,[15] for it was contradicted by a paper published by John Michell in *Philosophical Transactions* in 1767.[16]

News of the paper by Michell came to Herschel as an unwelcome surprise. Michell had been Professor of Geology at Cambridge, and his lengthy paper, "An Inquiry into the Possible Parallax and Magnitude of the Fixed Stars, from the Quantity of Light which They Afford Us, and the Particular Circumstances of Their Situation", was the most innovative contribution to the study of the stellar universe prior to Herschel's arrival on the scene. It was important enough for Herschel to have Caroline make a copy for him.[17]

When he read Michell's paper, he found, among much else, a remarkable – and highly original – probability argument. Michell's brilliant exposition deserves to be quoted at length:

> If now we compute, according to the principles above laid down, what the probability is, that no two stars, in the whole heavens, should have been within so small a distance from each other, as the two stars β Capricorni, to which I shall suppose about 230 stars only to be equal in brightness, we shall find it to be about 80 to 1.
>
> For an example, where more than two stars are concerned, we may take the six brightest of the Pleiades, and, supposing the whole number of those stars, which are equal in splendor to the faintest of these, to be about 1500, we shall find the odds to be near 500000 to 1, that no six stars, out of that number, scattered at random, in the whole heavens, would be within so small a distance from each other, as the Pleiades are . . .

> We may from hence, therefore, with the highest probability conclude (the odds against the contrary opinion being many million millions to one) that the stars are really collected together in clusters in some places, where they form a kind of systems, whilst in others there are either few or none of them, to whatever cause this may be owing, whether to their mutual gravitation, or to some other law or appointment of the Creator. And the natural conclusion from hence is, that it is highly probable in particular, and next to a certainty in general, that such double stars, &c. as appear to consist of two or more stars placed very near together, do really consist of stars placed near together, and under the influence of some general law, whenever the probability is very great, that there would not have been any such stars so near together, if all those, that are not less bright than themselves, had been scattered at random through the whole heavens.[18]

Michell had had no difficulties with what would later become Herschel's first postulate:

> And in the meantime, till this [annual] parallax be found . . . we can only suppose them [the stars], at a medium, to be equal in size to the Sun, this being the best means, which we have at present of forming some probable conjecture concerning the extent of the visible universe.[19]

The second postulate was another matter. When it came to classifying the stars by their appearance, Michell wanted more than the crude divisions into magnitudes inherited from Ptolemy, so that the stars "may be ranked with precision both according to their respective brightness, and the exact degree of it". He had made "some gross experiments" and he thought Sirius was from 400 to 1,000 times brighter than those of the sixth magnitude – not $6 \times 6 = 36$ as Herschel's second postulate proposed.[20]

Those involved had no inkling that in his parallax paper Herschel was following in the footsteps of no less a person than Isaac Newton; nor could they, as the manuscripts in question were unknown until recently. It was only after the publication of *Principia* in 1687 that Newton at last addressed the cosmological problem of how it was that the stars apparently remained 'fixed', motionless, despite their being free to move. Gravity, he claimed, was a universal force, and forces result in movements – yet (or so it seemed at the time) every single star was still in the place recorded for it in Antiquity. Newton's response was that Providence had distributed the stars throughout space with a high degree of uniformity, so that each star was pulled more-or-less equally in all directions by the surrounding stars, and therefore remained at rest. In defence of this claim of uniformity, Newton had compared predictions from his geometrical model with the numbers of stars listed in the catalogues, and for this he had made exactly the same assumptions as Herschel was to do.[21]

Today we define magnitudes so that, other things being equal, a sixth-magnitude star is ten times further than one of first magnitude. Using what would one day be Herschel's second postulate, Newton found that a consequence was that the stars – bizarrely – seemed to become increasingly dense as one moved outwards from the Sun; and so he had modified the postulate appropriately. Remarkably, when Newton's model of the universe was discussed by Edmond Halley in 1721, Halley advanced the very value that has become the modern definition.[22]

In a Postscript to his catalogue of double stars, Herschel thanks Banks for supplying him with a copy of a catalogue of double stars that had been published by Christian Mayer of Mannheim[23] (happily he had found that there was limited overlap with his own catalogue). Mayer, like Michell, thought that some doubles were companion stars in space; but Herschel was too committed to his current programme to be diverted by a mere argument from probability. He concludes his Postscript by saying that "in my opinion, it is much too soon to form any theories of small stars revolving round large ones".

His search for doubles had started off as a contribution to the determination of annual parallax but by now it had become an end in itself. And so, as soon as the second review was completed around Christmas 1781, he immediately embarked on a third, now using a higher magnitude and measuring the relative positions of the two stars more exactly. This took him until 26 September 1783, and resulted in another catalogue, this time of 434 objects.[24]

Michell (now the rector of a parish in the north of England) had recently sent for publication a second paper, praising Herschel's work but reminding readers of his own earlier paper, where he had argued that most double stars must be physical companions. Accordingly, "it is not improbable, that a few years may inform us, that some of the great number of double, triple, stars, etc., which have been observed by Mr. Herschel, are systems of bodies revolving about each other".[25]

Michell was right. Around the turn of the century Herschel re-examined several of his double stars, and he found that in some of them the relative positions of the two stars had indeed changed over the past two decades in a way that could only be explained by their rotation around each other.[26] But by happy chance there was one double for which he had more information: Castor, one of the most famous of double stars, which had been recognized as such by G. D. Cassini as long ago as 1678. Herschel had first measured the angle of the line joining the two components in 1779, and he remeasured the angle three times in the 1790s and then, more purposefully, from March 1800 onwards. By a lucky chance he mentioned what he was doing to Nevil Maskelyne, who recalled that Bradley had told him that in 1759 the line joining the components of Castor was parallel to the line joining Castor to its twin Pollux.[27] The methodical Maskelyne had made a written note of this information, and he managed to locate the note among his papers.[28] Herschel now had evidence of the change of angle of the two components stretching over nearly half-a-century. In a paper published in 1803 he estimated the period of revolution of the components to be "342 years and two months", astonishingly close to the modern value of 350 years.[29]

For the other doubles, his evidence was limited to changes that had taken place over the last twenty years or so. His outlook had changed in the interval. Gravity and its consequences had become central to his thinking, and he now welcomed further evidence that an attractive force was at work among the stars. It was in 1802 that he first announced that Michell's probability argument was well-founded, in the remarks that preceded his third (and final) catalogue of nebulae and clusters. Using fully justified italics to highlight his claim, he wrote of his recent observations of double stars that "*many of them have actually changed their situation with regard to each other, in a progressive course, denoting a periodical revolution round each other . . .*" (below, p. 163).[30] This was the most that could be shown by the two sets of observations at his disposal, made two decades apart; evidence that this force was indeed Newtonian gravity would require further observations a generation later. When this was confirmed, after Herschel's death, his son John thought the matter so significant that he had the inscription above his father's tomb amended to read: "He laid open to the eyes and intelligence of astronomers the vast gyrations of double stars."[31]

Why do some stars vary in brightness?

The ancients had believed the stars to be 'fixed', unchanging, not only in position but in brightness; and the medievals saw no reason to dispute this. It happens that none of the most prominent stars in the sky varies to a significant extent, and few of the lesser stars visible with the naked eye vary by more than a couple of magnitudes. And so it required the dramatic appearance in 1572 of what we classify as a supernova to capture the attention of the handful of observers of the time. Tycho Brahe was able to demonstrate that the object was indeed celestial, and from then on astronomers were alert to the possibility of changes in the heavens.[32]

In 1604 a second star flared up, and there were reports of others. Then, in 1638, at the time of an eclipse, the Frisian astronomer Johannes Phocylides Holwarda noticed in the constellation of the Whale a star of better than third magnitude that was not listed by Ptolemy or by Tycho and which therefore he thought must be new.[33] But no sooner had his long but routine account of its discovery and eventual disappearance been printed off, than the star re-appeared. In the years that followed the star continued to puzzle observers, but in 1667 Ismael Boulliau was able to announce that it reached a peak of brightness every 333 days.[34] For the first time a variable star – Mira Ceti, the Wonderful Star in the Whale, as it had become known – was lawlike in its behaviour at least to the extent that its maxima could be predicted. Bouillau also offered a physical explanation of such variations, based on the known behaviour of sunspots. The Sun with its spots rotated, and a rotating star with similar but much larger dark patches would regularly appear of reduced brightness when its rotation brought the patches to the side facing the observer. But sunspots also varied, and comparable variations in the dark patches could explain any irregular changes in brightness.

This triggered a search for variable stars, one that was – or at least seemed – productive. But alleged discoveries were difficult to challenge, and any variations, cyclic or otherwise, could easily be explained on the 'dark patches' hypothesis. And so it was that in the later years of the seventeenth century such 'discoveries' became commonplace,[35] with the result that eventually the subject fell into disrepute.

And there it remained for most of the eighteenth century. Then, in 1780, the young Edward Pigott, who had earlier visited Bath for some weeks and observed alongside Herschel,[36] moved with his father Nathaniel to York. Nathaniel was a wealthy amateur astronomer with traditional interests, but Edward preferred to pursue a subject that did not rely on fixed instruments: variable stars.[37] In the autumn of 1781 the younger Pigott began to comb the seventeenth-century literature for records of variations, and to observe the stars in question for himself. Within weeks he made the acquaintance of a teenage neighbour, John Goodricke, who was a deaf-mute. Before long the two formed a partnership to monitor variable stars: each would observe from his own home, and they would meet up next day to compare notes.

The following autumn one of the stars that they were regularly observing was Algol, or β Persei, a bright, second-magnitude star whose occasional dip in brightness had been noticed in 1667 by Geminiano Montanari. On 7 November Goodricke confirmed that Algol was second magnitude as usual, but five days later he was astonished to find it was now fourth magnitude. So rapid a drop in brightness had never been seen before.

The two friends now kept careful watch on Algol, and on 28 December both saw the star increase in brightness from fourth to second magnitude before their very eyes. It occurred to Pigott that Algol might have a satellite that regularly eclipsed its parent star, and the two observers set out to establish the period of the supposed satellite's orbit. This eventually proved to be less than three days, a truly astonishing result. Pigott generously allowed his handicapped friend to take all the credit for the eclipse theory in the resulting publication, but there the eighteen-year-old hedged his bets by offering the traditional 'dark patches' as an equally valid alternative.[38]

The eclipse theory – which we know to be correct – predicted not only that the drops in brightness would occur at regular intervals, but also that what we would call the light curve would be perfectly symmetric about the minimum point. Unlike the dark patches theory, which could explain almost anything and so was difficult to disprove, the eclipse theory made very specific predictions, and its authors eventually came to abandon it. Perhaps they were misled by changes in the seeing conditions, or perhaps they were influenced by the three other short-period variables they went on to discover, none of whose variations could be explained by simple eclipses.

Although Herschel was well aware of the hypothetical eclipses – and indeed was among the first to confirm Algol's frequent drops in brightness[39] – he always found the dark patches explanation to be adequate. As he wrote in 1796,

> Dark spots, or large portions of the surface, less luminous than the rest, turned alternately in certain directions, either towards or from us, will account for all the phaenomena of periodical changes in the lustre of the stars, so satisfactorily, that we certainly need not look out for any other cause.[40]

He did however make it possible, once and for all, to determine whether or not a given star was in fact variable. Pigott and Goodricke had compiled simple lists of neighbouring stars that were carefully arranged in order of diminishing brightness, so that a significant variation in any one of the stars would reveal itself by disturbing the sequence. Herschel did much the same, but on an heroic scale: in the closing years of the century he undertook a mammoth programme of observations in which he left to posterity four "Catalogues of the Comparative Brightness of the Stars", comprising in all nearly three thousand stars, and a treasure trove for future astronomers.[41]

The solar system's voyage through space

In almost every investigation he undertook, Herschel secured the high ground for himself by basing his arguments on his own unrivalled experience as an observer – with just one exception, where he relied entirely on data available to all.

As we have seen, when the eighteenth century dawned, every single star still appeared to be ‘fixed’ relative to the other stars (they lie at immense distances from us, and so their individual movements are almost imperceptible). This situation changed in 1718, when Edmond Halley noticed that three bright stars were no longer in the places assigned to them in Antiquity (they had ‘proper motions’).[42] Clearly, then, stars moved. The Sun was a star, so in which way was it moving?

The problem was how to distinguish the proper motions that were merely apparent and resulted from the voyage of the solar system through space (thus giving clues to its direction), from those that belonged to the individual stars themselves. As Bradley wrote in 1747:

> If our own Solar System be conceived to change its Place, with respect to Absolute Space; this might, in Process of Time, occasion an apparent Change in the angular Distances of the fixed Stars; and in such a Case, the Places of the nearest Stars being more affected, than of those that are very remote; their relative Positions might seem to alter; tho’ the Stars themselves were really immoveable. And on the other Hand, if our own System be at Rest, and any of the Stars really in Motion, this might likewise vary their apparent Positions; and the more so, the nearer they are to us, or the swifter their Motions are, or the more proper the Direction of the Motion is, to be rendered perceptible by us.[43]

Tobias Mayer of Göttingen explained how to identify the direction in which the solar system is travelling in a lecture he gave in 1760:

> For if the Sun and all the planets together with it, including our own home the Earth, were moving straight towards some region, all the stars that appear in that region would seem to be gradually separating from each other one by one, and those that are in the opposite part of the sky would seem to be joining up; just as when you are walking through a wood the trees that are in front of you seem to be separated and those that are behind you seem to be joined together.[44]

Unfortunately, as Mayer explained, it was hard at that time to tell which ‘proper motions’ were authentic and which spurious. Proper motions were (and are) determined by comparing the current position of a distant star against its recorded position at some date in the past; and the longer the time-span the better – other things being equal. But other things were not equal, for the further one went back into the past, the greater the inaccuracies of position measurements were likely to be. Indeed, it was only after Bradley’s announcement of aberration in 1729 and of the nutation of the Earth’s axis in 1747 that truly accurate stellar positions had become possible.

Mayer had compared the modern positions of some eighty stars against their positions recorded half-a-century earlier. For most of these stars the differences were no doubt due to instrumental error and ignorance of the various complications; but "there are no fewer than fifteen or twenty which, in the now sufficiently perceptible space of fifty years, were found to be endowed with their own proper motion".[45]

However, Mayer displayed a well-justified scepticism concerning the alleged changes in position of most of his stars:

> In order that it may more easily be judged to what extent both [pairs of coordinates] agree or disagree, I have noted the difference in separate columns. This indeed, as is clear, is so slight in [the case of] most of the stars that, within the limits of observational error, it appears that they are really fixed and subject to no movement that can be perceived by our sense organs. However, in some cases the differences are a little greater than can be attributed to the imperfection of the instruments. Consequently, it would appear to be not improbable that those which exceed 10 or 15 seconds of arc, and which on this account are distinguished from the rest, indicate a certain degree of movement. But in the case of Arcturus, Sirius, Procyon, Pollux, [α Aquilae, γ Piscium] and certain other stars, the difference between their former and present-day positions is so great, that there can be no doubt whatever concerning their movement.

For the first time, data on proper motions had been assembled in sufficient numbers for astronomers to investigate whether a pattern was present that revealed the motion of the solar system through space. But Mayer himself could find no such pattern: "... these motions of stars are not bound by a common law of this kind ... Perhaps the true and genuine reason for these movements will remain unknown for many centuries."[46]

The French astronomer Jerôme Lalande was equally pessimistic. In a memoir published in 1779 he remarked that it would only be when the Sun's motion had taken it perceptibly closer to one region of the sky and away from the opposite region, that this motion would be recognized.[47] But this did not stop Lalande from including the more convincing Mayer data in the supplementary vol. iv (1781) to the second edition of his *Astronomie*.[48] Each of the twelve stars he listed had a (putative) proper motion of 18 seconds or more in one or other coordinate. For completeness, Lalande also cited the supposed proper motion of each star in the other coordinate, although this did not imply that he had any confidence in the datum in question (rightly so, for three of them are in error not only in quantity but in sign). This was to have unforeseen consequences, for Herschel had an unsophisticated attitude to astronomical data: facts were facts as far as he was concerned (he would later put equal confidence in data whether they were recorded by him as a novice amateur in his first week of observing or as a highly experienced professional viewing with the largest reflector on Earth[49]).

Herschel had cut his astronomical teeth on Ferguson's *Astronomy* and there he would have read that "If our solar system changes its place with regard to absolute space, this must in process of time occasion an apparent change in the [angular] distances of the Stars from each other...".[50] In his 1781 paper "On the Parallax of the Fixed Stars" he explains how future changes in some of the double stars he has listed might eventually reveal the motion of the solar system.[51] In the same paper he also mentions a conversation with Professor Thomas Hornsby of Oxford "on the subject of the stars of the first magnitude that have a proper motion".[52] Herschel was clearly interested in the question of the direction in which the solar system is moving (what he termed the 'solar apex') even before he became a professional astronomer in the autumn of 1782; and he then lost no time in seeing whether he could identify a pattern in the known proper motions where others had failed.

He did not have access to Mayer's lecture, which was as well because he would have been tempted to take all his data at face value, but he did know of those of Mayer's data that had been selected and listed by Lalande. He also owned a copy of Nevil Maskelyne's *Astronomical Observations made at the Royal Observatory at Greenwich from the Year MDCCLXV to the year MDCCLXXIV* (London, 1776), and tucked away on p. iv of "Explanation and Use of the Tables" was a brief list of proper motions, seven in right ascension (R.A., the celestial equivalent of terrestrial longitude) and two in declination (dec., corresponding to terrestrial latitude): only nine quantities in total, but ones in which Herschel could place full confidence.

Herschel encourages us to think that he began his investigation with Maskelyne's data, and only subsequently moved on to those of Lalande. He had no trouble in understanding that if the solar system is moving towards a given apex, then as a result of the solar motion every star will appear to move (perceptibly or otherwise) along the great circle passing through the apex, the star, and the point in the sky opposite the apex (the 'antapex'), in the direction away from the apex. Figure 2.2 is a preliminary sketch by Herschel with these circles drawn for the apex that he finally derived (the sky is drawn as viewed from the 'outside', which suggests that Herschel was working with a celestial globe).

The proper motion data, however, were given in components in R.A. and dec. Fortunately, for a given apex the direction of expected changes in R.A. are simple enough: if we divide the sky into two halves by the great circle passing

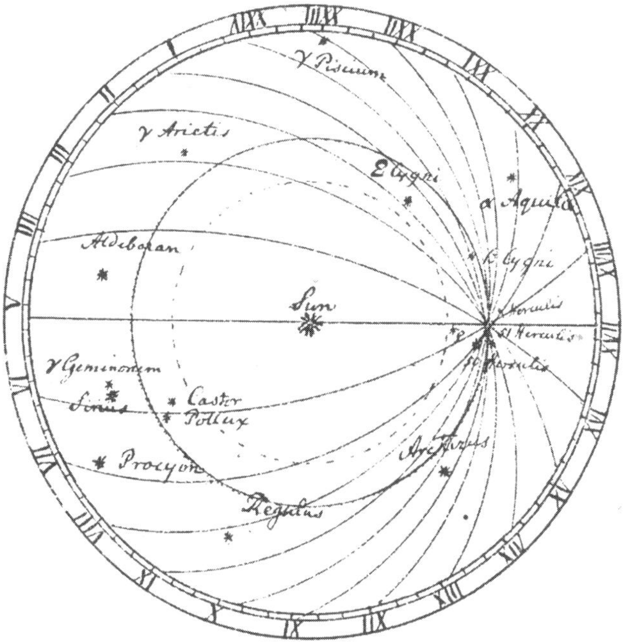

Fig. 2.2 Preliminary sketch by Herschel for Fig. 4 in his 1783 paper on the direction of the solar motion, showing the projection onto the equatorial plane of the stars whose proper motions are listed by Maskelyne or Lalande (Sirius is in the southern hemisphere). The curves radiating from λ Herculis are great circles along which stars will appear to move away from λ Herculis if that star is the direction towards which the solar system is moving. At this stage Herschel was also considering 50, 51 and ρ Herculis, for reasons discussed in the text. Note that the sky is reversed from that seen from Earth, suggesting that Herschel may have been working with a celestial globe. RAS W.7/4, f. 46, courtesy of the Royal Astronomical Society.

through the poles and the apex, then all the stars in one half will increase in R.A. while all those in the other half will decrease. Changes in declination, however, are more subtle and Herschel was to have difficulty with the geometry involved.

Happily, as we see in Fig. 2.3, every one of the directions of motion in R.A. of Maskelyne's seven stars is explained by the choice of an apex with R.A. between that of Arcturus and that of a star opposite to Sirius.

Figure 2.4 shows the corresponding data for the twelve Lalande stars. Six of the stars were also in Maskelyne's list, and happily both men agreed on the direction of their proper motions in R.A. As to the other six, three have changes so small that Lalande did not regard them as reliable – they might be no more than instrumental error, a mere 8 seconds in the change in R.A. of γ Geminorum in half-a-century, and a derisory 3 seconds in the case of β Cygni and Aldebaran.

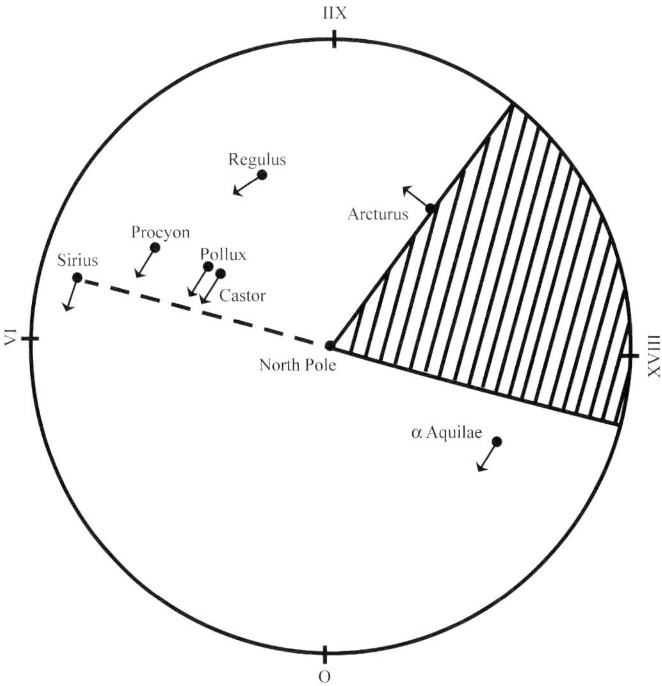
IIX
Regulus
Arcturus
Procyon
Sirius
Pollux
Castor
VI
North Pole
XVIII
α Aquilae
O

Fig. 2.3

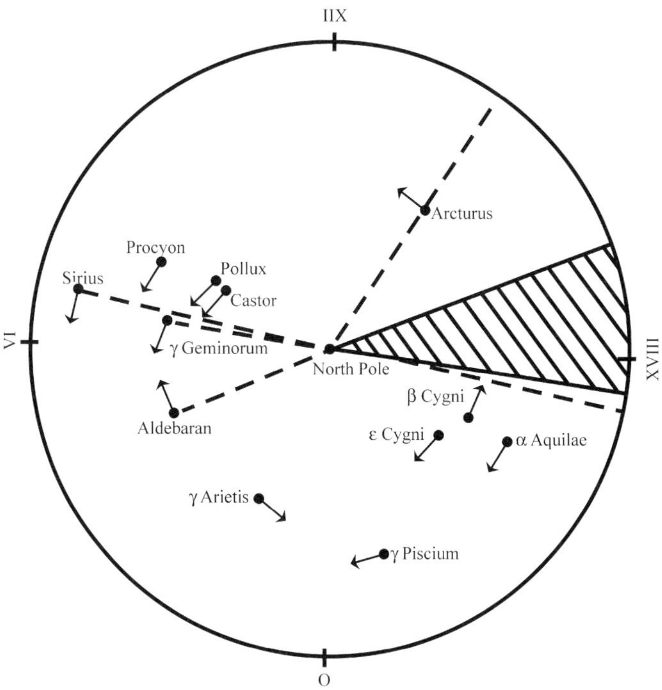
IIX
Arcturus
Procyon
Pollux
Sirius
Castor
VI
γ Geminorum
North Pole
XVIII
β Cygni
Aldebaran
ε Cygni
α Aquilae
γ Arietis
γ Piscium
O

Fig. 2.4

To Herschel, however, these tiny (alleged) changes were solid facts, to be accounted for (if possible) by his choice of the solar apex.

If we limit ourselves to the region defined by the Maskelyne motions, then we must accept γ Arietis and β Cygni as exceptions (in which case their changes in position resulting from the Sun's motion must be insufficient to overcome the (authentic?) changes in the opposite direction resulting from their individual motions). However, the change in Aldebaran – utterly negligible and therefore wrongly taken into account by Herschel – halves the region wherein the apex appears to lie (the change in γ Geminorum reduces it further, but by only a modest amount).

The pattern in declination is much more subtle and difficult to discern from the motions in declination of the Maskelyne and Lalande stars. Maskelyne gave only two such motions, both large and well-attested: Arcturus (in the northern hemisphere) was moving southwards, as was Sirius (in the southern hemisphere, but less than 17 degrees below the equator). Herschel perceived that within the region shaded in Fig. 2.4, an apex higher in the sky than Arcturus would be necessary.

Figure 2.5 indicates the directions of motion in declination of the Lalande stars in the northern hemisphere (that is, other than Sirius), although the change for γ Piscium is listed as only 7 seconds, that of α Aquilae as only 4 seconds, while for Castor Herschel takes seriously the absurd 'motion' of 1 second! We can see that with a selected apex very roughly between Arcturus and the North Pole, most of the stars are satisfactorily moving away from the apex. γ Piscium is a clear exception; but what are the implications of the northward motions of β and ε Cygni? Herschel thought that an apex within the shaded region in Fig. 2.4, and with declination north of Arcturus and south of β and ε Cygni, would produce the required southward motion of Arcturus and the northward motions of β and ε Cygni: that is, an apex within the shaded region in Fig. 2.5. Nearly central to this region is the star λ Herculis, which Herschel therefore provisionally fixed upon as an appropriate position for the solar apex. He accordingly, in February 1783, submitted to the Royal Society his paper "On the Proper Motion of the Sun and Solar System" (below, pp. 87–98),[53] in which he claimed that an apex near λ Herculis would satisfy all the Maskelyne and Lalande data with just three exceptions: γ Arietis and β Cygni in R.A., and γ Piscium in declination.

To satisfy the data with only three exceptions was, it seemed, a remarkable achievement. But with the apex at λ Herculis, the southward motions in declination of Sirius and Arcturus would be small, whereas Maskelyne and Lalande were agreed that the motions were large, and indeed that of Arcturus had been notoriously so for decades. Herschel felt these well-established southward

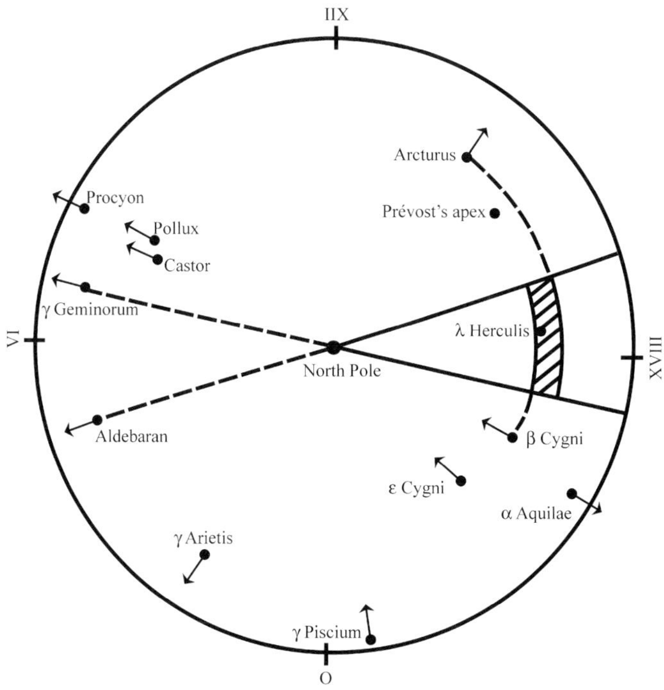

Fig. 2.5

motions of two very bright stars must be given great weight, and he therefore considered modifying the position of the proposed apex by moving it a little north, to near ρ Herculis (see Fig. 2.2), even though this would turn the motions in declination of β and ε Cygni into exceptions. But here he had made a geometrical blunder: they were exceptions even with the apex at λ Herculis, as Maskelyne told him a few days after the paper was read to the Royal Society:

> In addition to the remarks I sent you on your paper on the motion of the solar system, let me add, that you seem to have made a mistake in calculating the effect of the supposed motion towards λ Herculis on the declination of the stars; for by that motion β & ε Cygni will be carried a little southward, instead of northward as you suppose; and therefore contrary to the observations you adduce from M. DeLalande. I find too that upon your hypothesis the motions of Arcturus & Sirius southward will be very small, contrary to observation; nor can the matter be mended by altering the apex of the solar motion as you call it; as that would make the motions of the other stars to differ still more from the hypothesis than they do at present.[54]

Maskelyne was no hostile critic – "Your hypothesis struck me when I wrote the preface to my Tables & is probable" – but he knew of another star of the third magnitude "which has a considerable proper motion in contradiction to it".

The simplicity on which Herschel's progress had depended was rapidly being obscured. His friend, the amateur observer Alexander Aubert, contributed further to this by sending him a copy of Mayer's *Opera inedita*, with the 1760 lecture and the full list of eighty real or imaginary changes, and he did so on the very day when Herschel's paper was read to the Royal Society.[55] It would be impossible to cope with so many data by the simple graphical methods Herschel had so far used – and Mayer himself had ascribed the majority of the supposed changes to instrumental error.

Herschel was uncomfortable when the data themselves were suspect. And although he might change his own mind on occasion, he was difficult to budge once he had taken a stand. But the situation was now getting out of hand. He made a half-hearted attempt to check the existing results against the new data, and now rejected as "insensible" changes of less than 10 seconds even though such changes, especially in the R.A. of Aldebaran, had earlier been taken as evidence of decisive importance. And so he composed a Postscript (not reproduced) in which he claimed to check the additional Mayer changes of 10 seconds or more in either coordinate against an apex in the region of λ or ρ Herculis; and he produced a list of stars whose 'motions' were compatible with such an apex, and a list of those whose 'motions' were not. There are signs of special pleading, but in general Herschel was right to claim that, more often than not, the new data favoured an apex near λ or ρ Herculis.

Later the same year, the Swiss philosopher and physicist Pierre Prévost published another estimate of the solar apex based on the Mayer data.[56] He and Herschel agreed in the declination of the apex, but they differed by 30 degrees in R.A. In fact the Mayer data, properly considered, permit a choice of R.A. anywhere in the range 227 degrees (or even less) to 268 degrees (or more). Herschel seems to have selected λ Herculis (R.A. 260 degrees) for the apex because it was midway between the bounds defined by Aldebaran (R.A. 65 + 180 → 245) and γ Geminorum (96 + 180 → 276), and he was diverted from the equally satisfactory range from (say) 227 to 245 because he gave weight to the negligible difference of 3 seconds between the positions of Aldebaran as recorded by Römer in 1706 and by Mayer in 1756. Because of this error, it happens that Herschel's position for the apex is almost exactly that of modern astronomy, whereas Prévost's is nearly 30 degrees away in R.A.

Never one to let well alone, Herschel returned to the problem two decades later, and he published two closely related papers, the first in 1805 in which he recalculated the direction of the solar motion, and the second in 1806 in which he estimated its velocity.[57] By then the length of time over which reliable positions of stars had been available had grown significantly, and to oblige Herschel the meticulous Maskelyne had recently determined the proper motions of the 36 stars that he had chosen as fundamental for his observations, by comparing their positions in 1766 with their positions in 1802.[58] Unfortunately the data had now become too numerous to allow simple investigations of the kind that Herschel had undertaken in 1783, but he had his earlier value for the apex to guide him.

At the beginning of the 1805 paper, Herschel tells us that he plotted Maskelyne's 36 motions on a globe. If, ideally, all the listed values had been completely accurate and due entirely to the motion of the solar system, then these plots would have lain along great circles from the apex, through the star in question, to the antapex. In other words, all 36 circles would without exception have intersected each other at the apex and antapex.

In practice Maskelyne's data could not possibly be completely accurate, nor in all probability were the changes entirely the result of the motion of the solar system, but Herschel hoped that the points of intersections of the circles, two by two, might cluster sufficiently to indicate an approximate apex. And, sure enough, "I find no less than ten that are made by stars of the first magnitude, in a very limited part of the heavens, about the constellation of Hercules". These ten intersections were all associated with just six of the brightest stars, stars presumably close to us and so most likely to demonstrate apparent movements reflecting the motion of the Earth-based observer through space. Herschel accordingly selected an apex in Hercules, not far from the apex he had proposed in 1783.

The reader will not be surprised to learn that Herschel's choice of the six stars was open to question. In 1806 he would list the eight stars that came first in order of brightness as: Sirius, Arcturus, Capella, Lyra [Vega], Rigel and Betelgeuse (equal), Procyon and Aldebaran (equal). Yet the six that he selected to indicate the modified solar apex were Sirius, Arcturus, Capella, Lyra, Procyon and Aldebaran – for it was their intersections that so conveniently led to only a minor modification in his earlier apex.

In his 1806 paper Herschel turned to the question of the (relative) velocity through space of the solar system. In those days before astrophysics there was no information whatever available on the radial velocities of stars (their motions in the line-of-sight towards or away from the observer). As to their transverse velocities, these were given by their angular velocities (proper motions) multiplied by their distances.

And therein lay the problem, for the only clue that Herschel had to a star's distance was its apparent brightness, on the age-old but questionable assumption that the stars are physically uniform. Herschel knew of binary stars in which one member of the pair was much brighter than its companion, so the assumption was subject to exceptions, to say the least; but it was all he had to go on. In fact, the stars vary enormously in luminosity: Herschel reckoned Rigel, for example, to be 1.35 times farther than Sirius, whereas we know the correct figure to be more than 62.

Not surprisingly, Herschel now found himself in trouble. It seemed that when the solar motion had been allowed for, the residual velocities (due to the motions of the stars themselves) would be more numerous and even more closely clustered than the apparent ones. Yet the solar motion had been introduced in the first place for the very opposite purpose, to bring simplicity out of seeming complexity.

Herschel had already seen the problem when writing the 1805 paper:

> The chief object in view, when a solar motion was proposed to be deduced from observations of the proper motions of stars, was to take away many of these motions by investing the sun with a contrary one. But the solar motion, when its existence has been proved, will reveal so many concealed real motions, that we shall have a greater sum of them than it would be necessary to admit, if the sun were at rest.[59]

At its simplest, the difficulty in fact arose when a highly-luminous star appeared bright despite lying at a great distance. Being at a great distance, it had no discernible proper motion; appearing bright, Herschel supposed it to be near. If, therefore, it was near and yet had no motion *relative* to the Sun, it must be that it was keeping pace alongside the Sun as together they moved through space. In short, bright stars that had no known proper motions were (he thought) now revealed as members of a cluster of stars that included the Sun, all moving together through space. To do Herschel justice, he had foreseen this possibility back in 1783, when writing the postscript to his solar motion paper in the light of John Michell's probabilistic arguments for the existence of star systems: the study of proper motions, he argued, might offer clues as to "which stars may possibly belong to ours, and which to other systems".[60]

Given the resultant bewildering variety of relative velocities, Herschel assigned to the Sun a velocity that reduced as far as possible the inequalities in the velocities of the other stars. "His arguments are very ingenious", wrote Professor John Brinkley of Dublin University, "but there is necessarily so much hypothetical in them, that the mind cannot feel much confidence in his conclusion".[61]

In the years when Herschel was investigating the solar apex, there in fact existed manuscript records of the observations necessary to establish accurate proper motions in great numbers. James Bradley was the very man who had discovered both the aberration of light and the nutation of the Earth's axis, and there was no-one more alert than he to the factors that can adulterate the accuracy of observations of stellar positions. Bradley was appointed Astronomer Royal in 1742, and in 1750 he installed at Greenwich an 8-ft quadrant by John Bird. With it, he made his observations with meticulous attention to the recording of possible complications that might affect their accuracy. For many years after his death in 1762 these irreplaceable observations remained in manuscript, and even when they were published in 1798–1805 they were still in their raw state. Fortunately, a gifted and dedicated young German mathematician, Friedrich Wilhelm Bessel, set to work and extracted from Bradley's data the exact positions for 3,222 stars as they were in 1755, the midpoint of Bradley's observations. His volume, published in Königsberg in 1818, bore the proud title of *Fundamenta astronomiae pro anno 1755*; and now accurate positions of stars in 1755 could be compared with their current positions to reveal their proper motions.

Bessel however could find no pattern in these data that might indicate a solar apex. As Francis Baily put it in 1831:

> The additional information, however, which we have subsequently obtained on this subject, and the discovery of proper motions, in many of the stars, much greater in amount than any which he [Herschel] contemplated, do not appear to confirm this theory. And M. Bessel, who has investigated this subject [in 1818] with his usual care and ability, states that many of the stars indicate a point in the heavens very remote from the one alluded to by Sir William Herschel: and that, in fact, there is no one point in particular, towards which he can discover any such tendency.[62]

But the pendulum was about to swing. Just six years later, the German astronomer Friedrich Wilhelm August Argelander published an analysis of no fewer than 390 proper motions, a number large enough for him to divide the motions by size into three groups, which he treated independently.[63] Each group yielded a solar apex not far from that proposed by Herschel. Since then the reality of the solar motion and the constellation Hercules as the location of the apex has not been in doubt, and before long Herschel's 1805 paper was being compared with Newton's *Principia* "for sustained reflection and high philosophic thought".[64]

3

The riddle of the nebulae: (1) True nebulosity?

The prehistory of the mysterious milky patches in the sky known as nebulae goes back to Antiquity, but eighteenth-century writers often took their lead from the short paper that Edmond Halley published in 1715, "An Account of Several Nebulae or Lucid Spots Like Clouds".[1] Halley listed six nebulae: those known to us from Charles Messier's catalogue of 1781 as M11, M13, M22, M31 and M42, together with ω Centauri which is too far south to be visible from England. They are, Halley declared, "Light coming from an extraordinary great Space in the Ether; through which a lucid *Medium* is diffused, that shines with its own proper Lustre".

But a star system so distant that existing telescopes were not powerful enough to 'resolve' it into stars would also appear nebulous, and there could be little doubt that such systems existed. John Michell in his 1767 paper suggested that "Those stars, which are surrounded with nebulae, are probably only very great stars, which, on account of their superior magnitude, are singly visible, whilst the others, which compose the remaining parts of the same system, are so small as to escape our sight. And those nebulae, in which we can discover either none, or only a few stars, even with the assistance of the best telescopes, are probably systems, that are still more distant than the rest". And after considering how the Sun and its neighbours might appear to a spectator in the Pleiades, he remarked: "This may serve to show us that those nebulae, in which we cannot distinguish any stars, may yet reasonably be supposed to consist of stars, though too far distant to be singly visible."[2]

The question, therefore, was whether *all* nebulae were simply what Herschel was to term "clusters of stars in disguise",[3] as Michell thought, or whether Halley was right and some were formed of a "lucid medium", or "true nebulosity" as Herschel preferred to call it.

Herschel first read of nebulae in the brief chapter in Robert Smith's *Opticks* on "Telescopical discoveries in the Fixt Stars". Smith declared that the Milky Way was

> nothing else but a prodigious number of very minute stars, so close to one another that the naked eye can only perceive a whitish mixture of their faint lights. This was *Galileo*'s discovery, who found also that those faint stars, which Astronomers call *Nebulosae*, appeared through his telescope to be small clusters of very minute stars.[4]

In 1656 Christiaan Huygens had seen "a whitish cloud" in Orion's Sword, and Smith reproduced Huygens's sketch. According to Huygens, he tells readers,

> the three little stars very near one another (marked θ by *Bayer*), together with four more, shone out as it were through a whitish cloud, much brighter than the ambient sky: which being very black and serene caused that lucid part to appear like an aperture, that gave a prospect into a brighter region.[5]

An article in *Philosophical Transactions* in 1715, Smith continued, had reported five more of these "lucid spots", where it seemed that there was "perpetual uninterrupted day".

James Ferguson had more to say. He lists five of these six "lucid spots" under his heading with this name; of the Andromeda Nebula, our M31, he says intriguingly that it "is sometimes invisible", in which case it could not be a vast star system.[6] The sixth, the Orion Nebula, he lists under a separate heading, "Cloudy stars". These "look like dim Stars to the naked eye; but through a telescope they appear broad illuminated parts of the Sky; in some of which is one Star, in others more".

> But the most remarkable of all the cloudy Stars is that in the middle of *Orion's Sword*, where seven Stars, of which three are very close together, seem to shine through a cloud, very lucid near the middle, but faint and ill-defined about the edges. It looks like a gap in the sky, through which one may see (as it were) part of a much brighter region. Although most of these spaces are but a few minutes of a degree in breadth, yet since they are among the fixed Stars, they must be spaces larger than what is occupied by our Solar System; and in which there seems to be a perpetual uninterrupted day among numberless Worlds, which no human art ever can discover.[7]

Herschel was only a beginner in astronomy but this was a challenge that aroused his curiosity. Were some of the nebulae indeed lucid spots, perhaps with embedded stars? Or were they all like Galileo's *nebulosae*, "small clusters of very minute stars"?

Of five of Smith's lucid spots Herschel had little more than a simple list, and one of these was below his horizon; but of the sixth, the Orion Nebula, he had a crude sketch of how it had looked in the past, along with the account in Ferguson of its more recent appearance. Unlike professional astronomers of his day, the Bath violinist focused on the crucial fact that because a vast star system could not possibly alter shape in only a few years, a nebula that had demonstrably altered shape *must* be formed of true nebulosity. And so, on 1 March 1774, when he decided he had reached the stage when he might open his first observing book, one of the two objects Herschel observed was the Orion Nebula (Fig. 3.1). That night the seeing was poor, but it was better three nights later. And sure enough, the nebula looked quite different from the sketch in Smith: it had apparently altered shape in the interim. "From this we may infer that there are undoubtedly changes among the fixt stars, and perhaps from a careful observation of this Spot something might be concluded concerning the Nature of it."[8]

As a novice observer Herschel did not realize the extent to which changes in seeing conditions or in the instrumentation can affect the appearance of a nebula, and in any case the sketch by Huygens was crude. Herschel decided that, by observing the Orion Nebula at regular intervals, he might himself establish whether or not it was in fact altering shape. True, the ghostly form of the luminous spot was nearly impossible to record, and his skills as a draughtsman were limited; but if he kept meticulous records of the configuration and brightnesses of the embedded stars, and their relationship to the luminous spot, he might find evidence of change. In his remaining years in Bath, therefore, he returned time and again to the nebula. On 11 November 1776 he made elaborate notes of its appearance. On 25 January 1778 he repeated the exercise, and continued the following night; he records for example that embedded stars "6.2.1 make a straight line". On 7 and 25 February and 12 March he looked for but found no changes in the January configuration, and the same was true on 15 December, when stars 6, 2 and 1 again lay in a straight line. But on this occasion he felt sure "there is a visible alteration in the figure of the lucid part". He confirmed this on 7 October the following year: "the figure of the lucid part is much altered." On the same night he noted that "the line 6.2.1 is a little convex towards 5". On 5 December he decided "6.2.1 concave … the concave part turned to the south". On 22 January 1780, however, "6.2.1 instead of seeming concave towards the north appear convex"; but he could not be sure. On 19 and 26 February 1780 he looked again, and now began to measure the angles separating pairs of stars; and on 22 October 1781 we find him recording

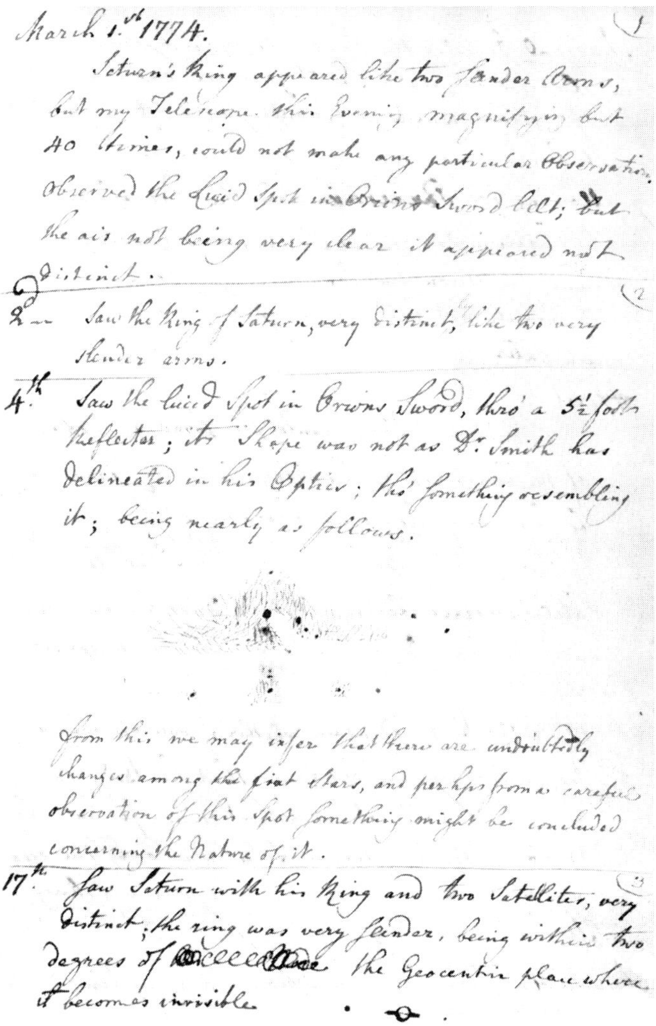

March 1st 1774.

Saturn's Ring appeared like two slender Arms, but my Telescope this Evening magnifying but 40 times, could not make any particular Observation. Observed the lucid Spot in Orion's Sword belt; but the air not being very clear it appeared not distinct.

2 — Saw the Ring of Saturn, very distinct, like two very slender arms.

4th Saw the lucid Spot in Orions Sword, thro' a 5½ foot Reflector; its Shape was not as Dr. Smith has delineated in his Optics; tho' something resembling it; being nearly as follows.

From this we may infer that there are undoubtedly changes among the fixt Stars, and perhaps from a careful observation of this Spot something might be concluded concerning the Nature of it.

17th Saw Saturn with his Ring and two Satellites, very distinct; the ring was very slender, being within two degrees of the Geocentric place where it becomes invisible.

Fig. 3.1 The opening page of Herschel's first observing book. On 4 March 1774 he observed and sketched the Orion Nebula, and comments that its shape was not as shown in the sketch (by Huygens in 1756) reproduced in Smith's *Opticks*. A nebula that altered shape in only a few years could not possibly be a distant (and vast) star system. RAS W.2/1.1, f. 1, courtesy of the Royal Astronomical Society.

the colours of the stars.[9] But by now he was emerging on the world-stage as the discoverer of a 'comet' that some thought might be a planet, and in the resulting upheaval it would be nearly a year before he could find time to think once more about nebulae.

Ferguson had given brief details of the locations of the four remaining nebulae listed by Halley that were visible from England. In his Bath years

Herschel was a busy musician; most of the time he could give to astronomy he spent improving his armoury of reflectors, and when he did have time to observe, the objects he examined were mostly nearer home. But he found occasions to view three of the four, and they provided him with evidence of the bewildering variety of forms to be found among the nebulae. M11 is a galactic star cluster, and Herschel had no problem in resolving it: 12 Sept 1779, "nebula seems to be a prodigious number of small stars surrounded with luster and glare"; 7 Aug 1780, "a great many stars"; 25 June 1781, "an amazing number of small stars"; 26 July 1781, "great number of Stars". M13, by contrast, is a globular cluster and could not be resolved by the small instruments then at Herschel's disposal: 22 Aug 1779, "without stars in it"; 29 Aug 1779, "there is no star in it". M31, the Andromeda Nebula, is of course a galaxy and far beyond the resolving power of any instrument of the Herschel era: 6 Aug 1780, "Nebula, in Andromeda, has no Star in it"; 27 July 1781, "no star visible".

Herschel also came across five other nebulae in his Bath years. On 2 August 1780 he noted that χ Persei (the galactic cluster NGC 884) is "An astonishing number of small stars". On 30 September 1780 he found near λ Aurigae "a very rich Spot", and this will be the galactic cluster NGC 1857. On 15 July 1781, "over β [Serpentarii] is a fine cluster of small stars", and this is the galactic cluster IC 4665 now in Ophiuchus. In addition, on 24 August 1780 he came across two nebulae with stars, one in Sagittarius and the other in Ophiuchus; but the locations he gives for them are vague.

For his first observations of the Orion Nebula Herschel had used a home-made Newtonian reflector of 5½-ft focal length. He was a man for whom 'more' and 'larger' were always 'better', and he did not need Smith to tell him that a telescope with greater aperture will bring into view faint objects that were previously invisible – and will allow higher magnifications to be brought to bear on objects already in view. On 1 May 1776 he used a 7-ft reflector (aperture 6 inches) for the first time (Fig. 3.2), and a few days later a 10-ft (aperture 8 inches). Within a couple of years the 7-ft and the 10-ft were to be mounted in simple, practical wooden stands of a design that he followed throughout his life.[10] The tube of the 20-ft with 12-inch mirrors that he completed in July 1776, however, was crudely slung by ropes from a vertical pole (Fig. 3.3); the unfortunate nighttime observer had to perch precariously alongside on the top of a huge ladder, at the risk of life and limb. Although Herschel devised a fine-tuning mechanism by which the observer could alter its vertical and lateral motions,[11] it must have been a nightmare to use – as he noted ruefully on 17 April 1777, "The air being changed and a high wind arose, I found it impossible to continue my obs. so long as I proposed . . . the uneasy posture and cold prevented farther observ:".[12] At Bath he used it very occasionally to observe Mars, Jupiter or Saturn

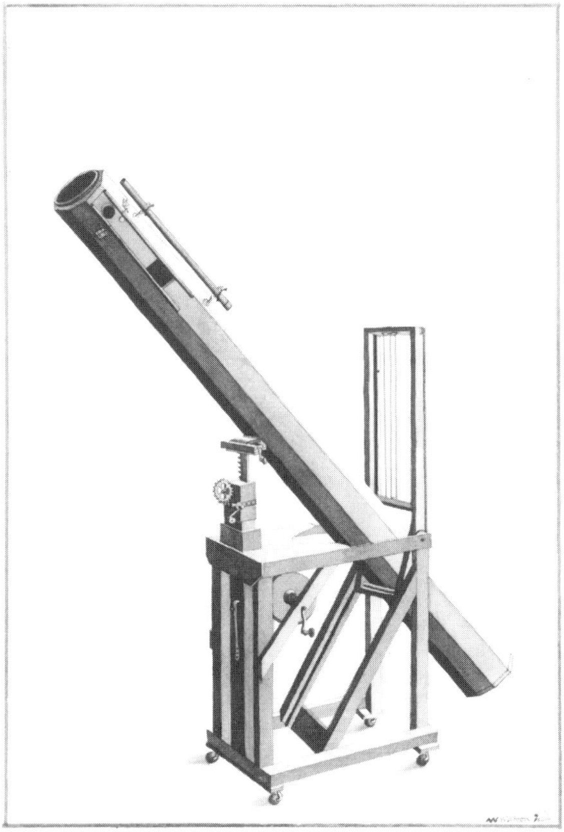

Fig. 3.2 In May 1776 Herschel made himself a 7-ft reflector with mirrors just over 6 inches in diameter. For two years he experimented with various mountings before arriving at the design shown in this sketch by William Watson, a design that Herschel was to follow throughout his career as a manufacturer of reflectors. In November 1778 he achieved a mirror that was the finest of its kind anywhere, and which bore fruit in his early catalogues of double stars and in his discovery of Uranus. RAS W.5/5, no. 3, courtesy of the Royal Astronomical Society.

(although never Lunarians or the mountains on the Moon), but there was no possibility of his sketching the layout of the stars of the Orion Nebula in the dark at the top of a 20-ft ladder.

The 10-ft with its modest 8-inch aperture was not an instrument that would satisfy the ambitions of any committed student of the construction of the heavens: James Short, the eminent maker who had died a few years before, used to offer for sale Gregorian reflectors with 18-inch mirrors.[13] But Herschel was currently preoccupied with earning his living as a musician, and when eventually he did have time to devote to systematic observations he wisely

Fig. 3.3 The 20-ft reflector with 12-inch mirrors that Herschel completed in July 1776, and which for seven years was the largest at his disposal. It was slung from a pole in the manner of the long refractors of the seventeenth century. But whereas with a refractor the eyepiece is at the bottom of the tube, in a Newtonian reflector it is near the top, and when observing near the zenith Herschel would be nearly 20-ft above the ground. From the sketch by William Watson, RAS W.5/5, no. 4, courtesy of the Royal Astronomical Society.

concentrated on the nearer stars, using the superlative mirror for the 7-ft that he completed in November 1778. But he hankered after a reflector with huge mirrors that would allow him to examine the fainter nebulae, and he may have been spurred into action by the news early in 1781 that the founder employed by John Michell had successfully cast a speculum disc of nearly 30 inches diameter. On 12 August 1780 Herschel had written to Joseph Priestley to ask about Michell's plans, and on 21 January 1781 Michell sent news of the casting direct to Herschel's Bath ally Dr William Watson Jr.[14] Michell alone of Herschel's English contemporaries was a dedicated student of the large-scale universe – although, curiously, he was to make no use of his great telescope to contribute to the subject, despite the promptings of friends.[15]

That same January Herschel embarked on the construction of a giant reflector, with mirrors that he eventually decided should be 3 ft in diameter – bigger even than Michell's. But whereas Michell had located a founder capable of work on this scale, Herschel ended up having to cast the mirrors himself in a foundry installed in the basement of his own home.[16] In August 1781 he twice tried to cast the great mirror, and twice failed; and for the time being he gave up the attempt. As 1782 opened, he still had grandiose plans for a large reflector,[17] but he found no opportunity to pursue the matter before August when he took up his appointment to be astronomer to the King.

King George was a serious student of astronomy – indeed, he had established Kew Observatory so that he might observe the 1769 transit of Venus – and as part of his duties Herschel was required on demand to transport his 7-ft to Windsor Castle from his home at Datchet. And so it was that at the end of August 1782 he showed the King "several nebulae, that in Andromeda [M31] and that in Scut. Sob. [M11]". By October Herschel had devised some sort of tripod support for his 10-ft, no doubt so that it too could be transported to the castle, and on the 12th he was able again to show M11 to the King – but this time with a mirror of nearly twice the 'light-gathering power'.[18] And so the King experienced for himself the benefit of larger mirrors.

The 20-ft was much too large to transport, but George was keen to view it, and so on 1 December the King made the journey to the Herschel home.[19] But the precarious observing ladder was no place for royalty. He was consoled with a view of Herschel's planet, now named the Georgium Sidus in honour of the King, through the 10-ft.[20]

The King did not observe again until 28 April, when Herschel took the 10-ft to the castle and again showed him his planet, along with the scattered cluster in Gemini, M35.[21] By this time Herschel was making progress with the construction of what was to prove one of the most successful telescopes of all time. It was another 20-ft, but its mirrors were to be 18 inches in diameter. This was only a modest improvement on the existing 20-ft, and far short of the 30-ft with 3-ft mirrors that he had attempted in Bath; but this time the technology was within easy grasp. Experience with the existing 20-ft had taught him the crucial importance of a convenient and safe mounting, and the new instrument was to be housed in a ladder structure that carried an observing platform complete with guard-rails (Fig. 3.4). From the platform Herschel would be able to manhandle the tube from side to side. Alternatively, if the instrument was to be kept facing exactly south (so acting as a transit instrument), he could use an observing chair securely attached to the ladder structure (Fig. 3.5). The 'large' 20-ft with its 18-inch mirrors was completed in October 1783, and by the close of the year Herschel had recruited Caroline to be his amanuensis in a two-decades-long observing campaign with the 20-ft that was to add an astonishing 2,500 nebulae to those previously known.[22] With the telescope facing due south, Herschel would be at the eyepiece scanning the sky as it rotated before his eyes, while Caroline was at a desk at a nearby window, ready to copy down his shouted descriptions of any nebulae that came into his field of view and to record their positions. "This will be a work of some years", he wrote to a friend in his first

Fig. 3.4 Herschel's 'large' 20-ft reflector with 18-inch mirrors, which came into service in the autumn of 1783. It is shown in its prime, from an engraving he published in 1794. The gallery enabled him to move the tube from side to side and this is what he did in his first, solo 'sweeps' for nebulae; the gallery would also allow him to track a celestial object for several minutes. However, in his sweeps for nebulae with Caroline as amanuensis that began around Christmas 1783 he found it simpler to keep the telescope facing due south and he could then replace the gallery with an observing chair (Fig. 3.5). In September 1786, however, he repeated earlier experiments in which he peered directly down the tube through an eyepiece at the edge of the aperture, rather than through an eyepiece mounted in the side of the tube (as in the 'Newtonian' configuration). This proved advantageous, for it allowed him to dispense with the secondary mirror and so avoid the loss of light that reflection in this mirror entailed. From then on he always used the 'front-view' or 'Herschelian' configuration in his larger reflectors, and to permit this the telescope had to be equipped with a gallery rather than an observing chair.

flush of enthusiasm, "but it is, to me, so far from laborious, that it is attended with the utmost delight".[23]

But the 18-inch mirrors had only one-quarter the light-gathering power of 3-ft mirrors such as he had tried and failed to cast in Bath, and at the start of 1784 the question of the existence or otherwise of true nebulosity was as open as ever. Unfortunately, Herschel's salary from the King was less than half his earnings as a Bath musician, and he had drawn on his savings to pay for the 'large' 20-ft. To build a reflector whose speculum mirror would dwarf even Michell's would be possible only with funding from King George, and by the summer of 1784 his campaign to secure this funding was well under way. That August we find

Fig. 3.5 Herschel's 'large' 20-ft reflector, shown at Datchet in this watercolour by Rev. Thomas Rackham with the observer (in broad daylight!) in the observing chair. When the chair was in use the telescope would be facing south and acting as a transit instrument, and the sky in the field of view would change slowly with the rotation of the Earth. Herschel Family Archives.

Watson writing to him to ask, "above all, has the King set you about the magnum opus, the great speculum?"[24] The omens were good, for Herschel had the enthusiastic backing of Sir Joseph Banks, president of the Royal Society and a man with the ear of the King; and George had seen for himself the benefits of large mirrors.

A transformation in Herschel's desk knowledge of nebulae had taken place back in December 1781. Until then he had known of the existence of only a handful of these mysterious objects – those listed by Smith and Ferguson, plus the very few others that he had happened upon by chance. But that month Watson had sent him a catalogue of no fewer than 70 nebulae and clusters of stars that the French comet-hunter Charles Messier had compiled.[25] Some had been discovered by Messier himself with his 3½-inch refractor, some by other observers, and Messier had assembled the list because at first glance nebulae looked like comets and the distractions were costing him valuable time.

When the catalogue arrived from Watson, Herschel had in his garden the 20- ft reflector whose 12-inch mirrors were large by Messier's modest standards; surely he would take the earliest possible opportunity to use the telescope on

Messier's nebulae and see whether the description of some of them as nebulae without stars, "nébuleuses sans étoiles", survived this more rigorous examination. And this, he tells us in his 1784 *Philosophical Transactions* paper on the construction of the heavens (below, pp. 99–112),[26] is exactly what he did:

> As soon as the [catalogue] came to my hands, I applied my . . . reflector of 12 inches aperture to them; and saw, with the greatest pleasure, that most of the nebulae, which I had an opportunity of examining in proper situations, yielded to the force of my light and power, and were resolved into stars. For instance, [he lists 29 Messier objects], all of which are said to be nebulae without stars, have either plainly appeared to be nothing but stars, or at least to contain stars, and to shew every other indication of consisting of them entirely (p. 101).

This is in fact the very opposite of what really happened (see Table 3.1). Between the arrival of the gift from Watson in December 1781, and his move to Datchet in August 1782, Herschel did not look at one single nebula: he had more urgent things on his mind, and Messier's catalogue he simply put on one side. Only after he was established as Court astronomer did Herschel – very occasionally – observe a nebula.

People tell lies for a reason. The aperture of his largest reflector being only 18 inches, Herschel was desperate for the King to fund a reflector with aperture 48 inches. He was endlessly frustrated with the modest light-gathering power currently at his disposal, as we see from his observing books: "more power & light will probably resolve it" (M79, 28 Sept 1783); "I think a great deal more of light and a much higher power would be of service" (M17, 31 July 1783); "This speculum has not light enough" (M31, 2 Aug 1783); "wants more light to attack it properly" (M32, 2 Aug 1783); "I want light" (M27, 2 Aug 1783); and so on.[27] The monster reflector would be very expensive and Herschel needed to conjure up the most persuasive justification. And so he invented a plausible untruth: the dramatic new insights that had (or more correctly, would have) resulted two years before, when (or rather, if) the objects examined by Messier with a 3½-inch refractor had been re-examined with the 12-inch reflector.

Even in the months after his arrival at Datchet, Herschel had devoted most of his observing time to his very productive search with the 7-ft for double stars (a search that was to continue until September 1783). As a result, even as late as the end of March 1783 he had seen only nine Messier objects that had been unknown to him in his Bath days, all of them viewed with the 7-ft. We are well informed on such matters because we have both his rough observing notes and the fair copy that Caroline wrote up next day; we have too the file she

Table 3.1 *The Messier nebulae Herschel claimed to have resolved (wholly or partially) with his 'small' 20-ft reflector in the winter of 1781/82, and the dates on which they were in fact observed and the instrument (7-ft, 10-ft or 'small' 20-ft) used.*

Messier no.	Type	Date observed	Reflector	Herschel's comments
2	glob. cl.	31 July 83	20	"I can count 18 or 20 of the stars"
5	glob. cl.	21 May 83	10	"All stars"
9	glob. cl.	3 May 83	10	"connection between Nebulas in general & stars in clusters"
10	glob. cl.	21 May 83	10	"A cluster of very small stars"
12	glob. cl.	21 May 83	10	"A cluster of close stars of different sizes"
13	glob. cl.	22 Aug 79	7	"without stars"
14	glob. cl.	23 July 83	20	"I can distinguish a few stars & have not the least doubt it consists intirely of them"
15	glob. cl.	31 May 83	10	"all fairly resolved into stars"
16	gal. cl.	30 July 83	20	"Large stars with small ones among them"*
19	glob. cl.	28 May 83	10	" resolved into stars"
22	glob. cl.	4 July 83	20	"all resolved into stars"
24	gal. cl.	2 Aug 83	20	"Considerable stars in great number"*
28	glob. cl.	4 July 83	20	"if the night were clearer I doubt not of its being stars that might be seen very distinctly"
30	glob. cl.	31 July 83	20	"consists of very small stars"
31	sp. gal.	2 Aug 83	7	"227 [mag] A strong suspicion of stars. This speculum has not light enough. I doubt not but 20 feet with confirm it. 460, suspicion still stronger"
37	gal. cl.	24 Aug 83	7	"many small stars"*
51	sp. gal.	20 Sep 83	20	"Most difficult to resolve; yet I no longer doubt"
52	gal. cl.	29 Aug 83	7	"all resolved"*
53	glob. cl.	30 May 83	10	"More than a suspicion of stars"
55	glob. cl.	30 July 83	20	"fairly resolved into very small stars"
56	glob. cl.	31 May 83	10	"All resolved into stars"
62	glob. cl.	28 May 83	10	"A strong suspicion amounting to almost a certainty of its consisting of stars tho' I can distinguish none"
65	sp. gal.	31 May 83	20	"almost a certainty of its being stars. There is however in both the nebulas [65 and 66] a certain whitishness left which may be nebulosity"
66	sp. gal.	31 May 83	20	"A strong suspicion of stars"
67	gal. cl.	26 Mar 83	7	"A cluster of stars"*
71	glob. cl.	30 May 83	10	"resolved into stars. I can count between 20 & 30 of them"
72	glob. cl.	28 Sep 83	10	"fairly resolved into very small stars"
[74	sp. gal.	20 Sep 83	20	misidentification]
92	glob. cl.	25 Aug 83	20	"A most beautiful sight. I can count 50 or 60 stars besides numberless that only distinguish themselves by twinkling. . ."

*Comment quoted *in extenso* in J. L. E. Dreyer (ed.), *The Scientific Papers of Sir William Herschel* (London, 1912), **2**, 651–660.

compiled in later years listing the Messier objects and the dates when her brother first observed them.[28] By the end of March a full fifteen months had elapsed since the arrival of the gift from Watson, and Herschel had still not examined a single Messier object with the 'small' 20-ft.

By then Caroline was finding nebulae in 'sweeps' with the little refractor her brother had given her, and this had opened Herschel's eyes to the great number of nebulae awaiting discovery. He impulsively began himself to sweep with a small refractor, before he accepted that nebulae were permanent objects in the night sky and deserved to be examined at leisure with the biggest telescope available.[29] The 'large' 20-ft was approaching completion (it came into operation in late October 1783), and during the summer Herschel indulged himself by giving whatever time he could spare from double stars to examining objects in the Messier list – or rather lists, for around the end of May he acquired a copy of Messier's third and final catalogue of 103 objects, the one we use today.[30] It was during these summer months of 1783 – and not the winter of 1781/82 – that Herschel viewed most of the Messier objects cited in his 1784 paper.[31] Once the 'large' 20-ft was commissioned, he embarked on his great campaign to sweep the whole of the visible sky for nebulae;[32] and from then on Messier objects took their turn with the rest.

Of the Messier objects that Herschel viewed from Datchet prior to the commissioning of the 'large' 20-ft, seventeen were declared by Messier to be clusters of stars and Herschel could do no more than confirm this: M4, M6, M7, M11, M18, M20, M21, M23, M25, M26, M29, M34, M35, M36, M38, M73, M103. With another five objects Herschel had to admit defeat: M17 (a gaseous emission nebula, seen in the 'small' 20-ft on 2 August 1783 as "A curious train of light. I cannot resolve it"); M32 (an elliptical galaxy, seen in the 7-ft the same night as "Like a nebulous star"); M43 (a gaseous emission nebula, seen in the 20-ft on 20 September 1783 as "nebulous"); M78 (a gaseous reflection nebula, seen in the 20-ft the same night as "2 large stars . . . with nebulosity"); and of course the Orion Nebula, M42. At Datchet Herschel abandoned his earlier caution in pronouncing on changes in the nebulosity of M42: on 31 Jan 1783 he recorded that "the nebulous part is quite different from what it was last year" (7-ft); on 20 Sept 1783 it "has evidently changed its shape since I saw it last" (20-ft); a week later he was writing of "the new division in the Nebula of Orion" (20-ft?); and the following night he declared, "Nebula in Orion is surprizingly changed" (7-ft).[33]

So what of the twenty-nine objects Messier saw as "nebulae without stars", that Herschel claimed to have resolved – in part, if not completely – into stars with the 'small' 20-ft in the winter of 1781/82? The last four of them were not even in Messier's second catalogue and therefore could not possibly have been examined by Herschel at the date claimed. As to the telescopes he in fact used, he saw just thirteen of the twenty-nine with the 20-ft, all during the summer of 1783. Five he viewed only with the 7-ft, and ten with the 10-ft. The inclusion of the remaining nebula, M13, must be a slip, for he had twice seen it from Bath in August 1779 as "without stars" and did not see it again until he was sweeping with the 'large' 20-ft in 1787. The citation of M74 is also erroneous, for the object – whatever it was (M71?) – is wrongly labelled in the observing notes.[34]

In his 1784 paper Herschel also listed nine Messier objects (M1, M3, M27, M33, M57, M79, M81, M82 and M101) that in one or other of his reflectors "shewed a mottled kind of nebulosity, which I shall call resolvable" (pp. 100–101). He currently believed that some of the nebulae he was viewing were truly nebulous, continuous clouds of luminous fluid, while others were formed of individual stars. How then was one to tell the difference? He sometimes saw in the sky a smooth or milky nebulosity, but sometimes the nebulosity was "mottled". Very plausibly, he took the milky nebulosity to indicate the presence of luminous fluid, and the mottled nebulosity to be star clusters on the verge of resolution into their component stars.

Herschel had long understood that there were two main tests for whether or not a nebula was in fact a star cluster disguised by distance. If a nebula altered shape, as (it seemed) had the Orion Nebula, it could not be a vast star cluster: there was therefore at least one undisputed example of true nebulosity. If on the other hand a nebula was resolved into stars when examined with a telescope of larger aperture than hitherto, then clearly it was a star cluster.

But to resolve one nebula, or even many, could never prove that *all* nebulae without exception were star clusters, for there seemed no limit to the number of nebulae in the sky. However, this logical point does not make due allowance for human nature: decades later, in 1845, when Lord Rosse's 'Leviathan of Parsonstown' seemed to show that the Orion Nebula was a star cluster after all, many astronomers would take this single 'resolution' to be demonstration that all nebulae were star clusters.[35] Herschel's 'small' 20-ft was currently resolving numbers of Messier objects into stars (even if the resolution was not always complete); and many more seemed on the verge of resolution, or would surely be resolved on a finer night. One could cite innumerable examples of this from his notes. For example:

> I see several stars in it & make no doubt, a higher power and more light would resolve it into stars. This seems to be a good Nebula for the purpose of establishing the connection between nebulas in general & stars in clusters [commenting on the globular cluster M9, 3 May 1783].

> . . . resolved into stars i.e. I can count 5 or 6; & all the rest of the light appears mottled like other nebulas when not sufficiently

> magnified and illuminated to shew the stars [on the globular cluster M19, 28 May 1783].
>
> Plainly resolved into very small stars. It is a difficult step i.e. if we divide the transition from the Pleiades down to the Nebula in Orion into six steps this is perhaps the 4th towards the real nebulas. The stars in this seem to be of two different sizes for I perceive 3 or 4 very visible ones branching out towards the north and several more exceedingly small at the sides. Towards the south in one place the light is very intense, but has all the appearance of crowded stars; so that there remains no doubt of the whole being stars [on the globular cluster M30, 21 August 1783].
>
> This & the 51st are both so far removed from the appearance of stars that it is the next step to not being able to resolve them. My new 20ft will probably render it easy [on the spiral nebulae M101 and M51, 20 September 1783].

Herschel several times imagines these six 'steps', with a simple cluster as Step 1 and the Orion and other "real" nebulae as Step 6. But it is difficult to resist the impression that he sees Step 1 as corresponding to the resolving power of the unaided human eye; Step 2, to that of a small telescope such as Messier's; Steps 3 and 4, to that of his present equipment; and Step 5, perhaps to that of the 'large' 20-ft soon to be commissioned. And what of Step 6? Would the giant reflector that the King might fund show the objects in Step 6 also to be clusters of stars? Revealingly, in a manuscript setting out ideas that would mature in his 1784 and 1785 papers on the construction of the heavens and which must date to 1783 or early 1784, he writes: "... the Nebulae are not any sort of vapours extended before the vacant spaces but are a collected body either of luminous matter or *most probably of stars* [italics supplied]".[36] Yet surely the Orion Nebula had altered shape, and so this at least must be a nearby cloud of true nebulosity. And so, when Herschel sent his first paper on the construction of the heavens to the Royal Society in April 1784,[37] he was still contrasting "nebulosity of the milky kind, like that wonderful, inexplicable phaenomenon about θ Orionis", with the nebulae that "shine with a fainter, mottled kind of light, which denotes their being resolvable into stars" (p. 106).

Even Herschel's 'large' 20-ft had mirrors of only 18 inches diameter, far inferior to the 30-inch mirror that Michell had successfully cast three years earlier. Herschel was desperate to build himself a reflector with unrivalled 'light gathering power', with which to advance the study of the nebulae, and his 1784 paper – completed only a few weeks after sweeps with the new 20-ft had begun in earnest – is an opening shot in the campaign. Its first purpose is to illustrate the striking insights that the modest advance from 12 to 18 inches has already made possible, to whet the royal appetite for the revelations that would surely follow the construction of a reflector that would dwarf all others on the planet. And so Herschel opens his paper by describing his fine new 20-ft, after which he enumerates some of the marvels it has enabled him to see. The Milky Way now completely resolved into stars, something that "my former telescopes had not light enough to effect" (p. 100); indeed, in a single hour, over 50,000 stars might pass before his eyes (p. 101). Many of Messier's supposed "nebulae without stars" had long since (he claims) been resolved by his 12-inch; and the process is continuing. When he examines with the 18-inch a specimen of the nebulae discovered by Messier or by his colleague Pierre Méchain, he finds that the French astronomers "saw only the more luminous part of their nebulae" (p. 102). And in only a few weeks he has increased nearly sixfold the number of nebulae known to astronomy.

The second purpose of the paper is to explore the three-dimensional structure of the Milky Way. In 1610 Galileo's telescope[38] had confirmed the speculation of the Greek philosophers Anaxagoras and Democritus[39] that the Milky Way is formed of innumerable stars; but for well over a hundred years no one thought to investigate the three-dimensional shape of the system of stars that gives rise to this appearance. It was not until 1750 that a cosmological speculator, Thomas Wright of Durham, attempted an explanation of sorts.[40] According to Wright, the Sun is one of the stars that occupy a spherical shell of space that surrounds a Divine Centre (Fig. 3.6). The radius of the shell is vast, and the stars actually visible to us form only a tiny segment of the whole. Indeed, so vast is the radius that for practical purposes the visible stars may be imagined as though they belonged to a system bounded by parallel planes, even though this is not strictly the case.

At some time in his career – probably when he was in later life – Herschel acquired a copy of Wright's fanciful book, *An Original Theory or New Hypothesis of the Universe*; but we find in it only a single marginal comment, which suggests he found it of little interest.[41] But it is just possible that, as a young musician in the north of England, Herschel heard mention of Wright's ideas, and that later there surfaced in his mind the essential insight: that the Milky Way effect will result if the Sun is one star in a layer or 'stratum' of stars. For if the Earth-based observer looks out in directions within the plane of the layer, he will see innumerable stars, near and far, whose light will merge to give a milky effect; whereas if he looks outwards and away from the layer, he will see only nearby (and therefore bright) stars before his gaze reaches into empty space.

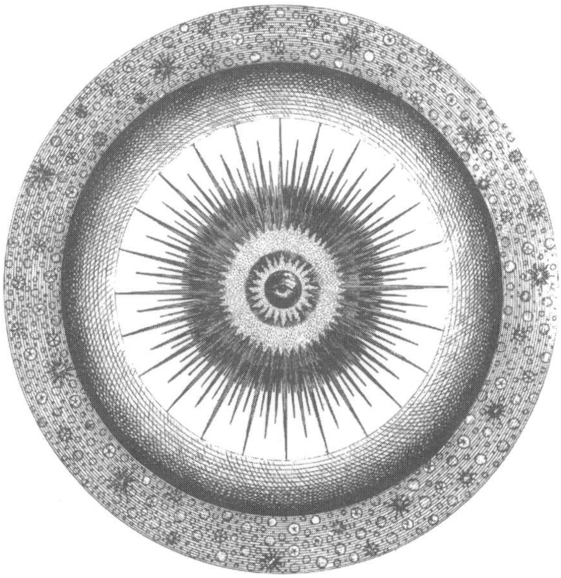

Fig. 3.6 Thomas Wright's conception of the spherical star-system to which the Sun belongs is shown here in cross-section. The system, which surrounds the Eye of Providence, is relatively thin but its radius is so vast that the stars *actually visible to us* belong to a layer not very different from what Herschel would later term 'a stratum'. From Wright's *An Original Theory or New Hypothesis of the Universe* (London, 1750).

In his 1784 paper Herschel declares that the Milky Way "undoubtedly is nothing but a stratum of fixed stars" (p. 106), possibly with "a branch, or smaller stratum", to account for the observed bifurcation in the Milky Way. Impressively, he goes on to outline a method by which the location of the Sun in the stratum might be established. It involves a pioneering exercise in stellar statistics, indeed the first of its kind in history. "I call it *Gaging the Heavens*, or the *Star-Gage*. It consists in repeatedly taking the number of stars in ten fields of view of my reflector very near each other" (p. 108), adding them and taking the average. He gives a brief illustration of this to show "that the number of stars increases very fast as we approach the Via Lactea" (p. 108). The number of stars in his field of view will provide a clue to the distance from us to the border of the stratum in that direction, provided that within the stratum the stars are distributed with fair uniformity. He accepts that the appearance of the Milky Way will be accounted for if the Sun is in the middle of a ring of stars, but he rejects this possibility on the basis of what today we term the Copernican Principle, that there is no reason to assign "this singular distinction" to ourselves (p. 108).

Given that the stars are indeed distributed with fair uniformity, then the larger the number of stars in a given direction, the greater the distance to the

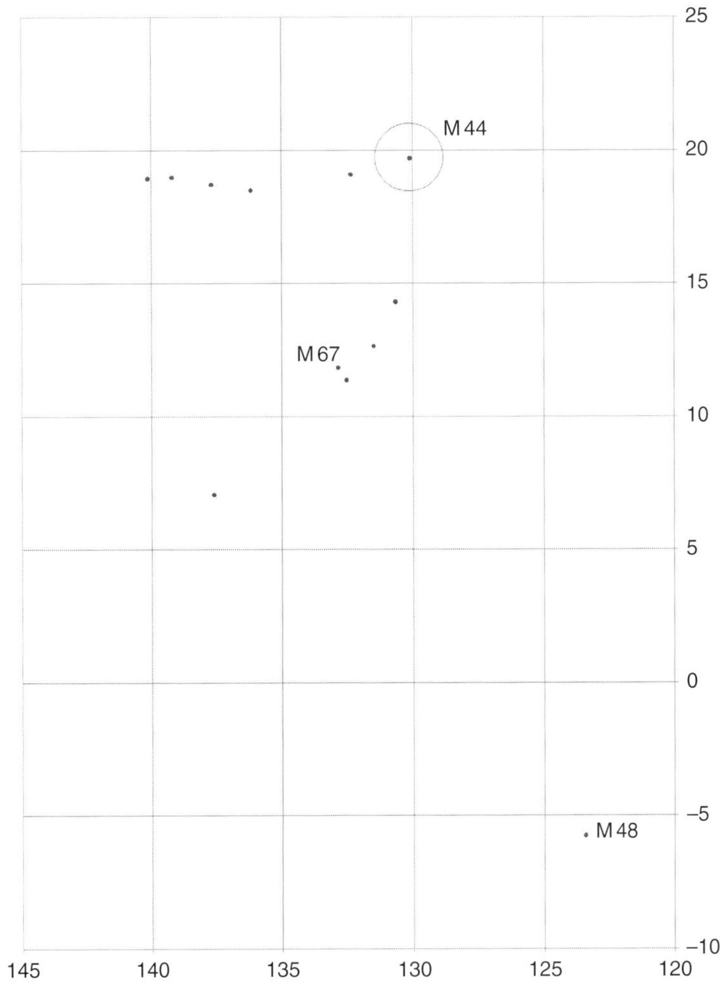

Fig. 3.7 Herschel's "Stratum of Cancer", covering parts of the constellations Cancer and Hydra. This diagram is based on all the 13 objects known to him when he submitted the 1784 paper. Most prominent are the open clusters M44 (Praesepe, visible to the naked eye), M48 and M67. Diagram by Wolfgang Steinicke.

border in that direction. He investigates in some detail the general structure of the compound stratum that makes up the Galaxy, and he claims that its greatest concentration of stars is in the direction that he has previously determined for the direction of the solar motion. The attractive force of this concentration of stars (a force that is no doubt gravitational, although he cannot be sure) would provide a physical cause for this motion (p. 109).

The declared subject of the paper is the "various *nebulous and sidereal strata* (to borrow a term from the natural historian)" (p. 100), and it is Herschel's

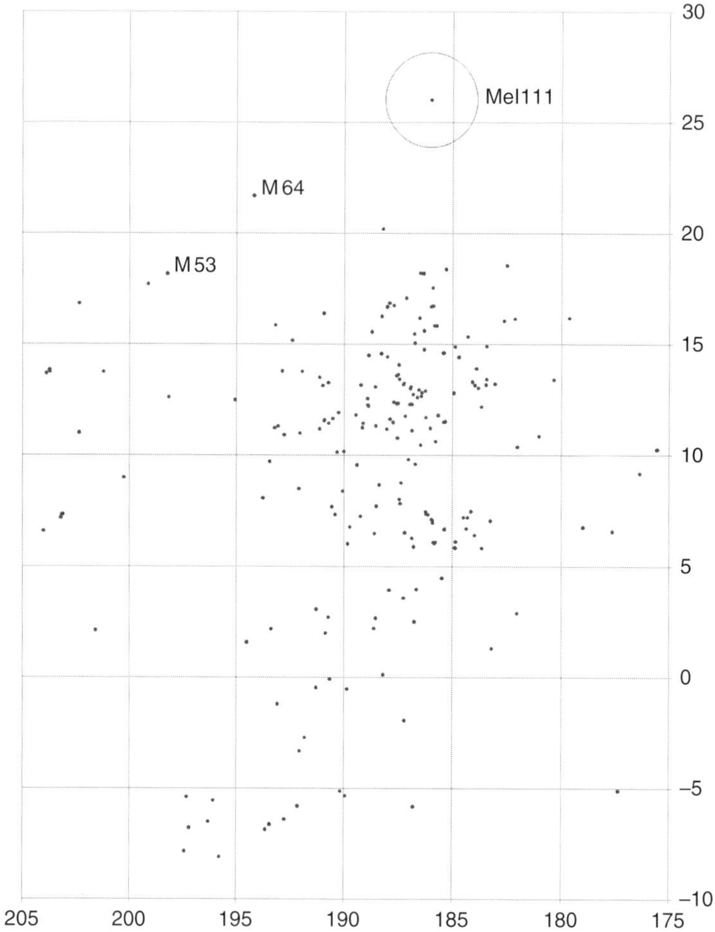

Fig. 3.8 Herschel's "Stratum of Coma Berenices" covers parts of the constellations Coma Berenices, Virgo and Ursa Major. The diagram is based on all the 209 objects known to him when he submitted the 1784 paper, the largest being the Coma Berenices cluster (Melotte 111, visible to the naked eye). Most of the objects are galaxies, belonging to the Virgo cluster of galaxies, and there are numerous Messier objects (not labelled). M53 is a globular cluster and M64 a nearby galaxy. Diagram by Wolfgang Steinicke.

discussion of nebulous strata that is least satisfactory. Eager to generalize from the stratum of the Milky Way, and already convinced that the destructive power of gravity is at work in the cosmos, Herschel sees the nebulae he has encountered as "arranged into strata" (p. 106) which have already partly fragmented. The concentrations of nebulae are real enough, but his interpretation is forced. One such stratum he thinks begins with the Praesepe cluster (M44) in Cancer, and extends away through the cluster M67 towards the head of Hydra (Fig. 3.7). Another begins with the stars of the constellation of Coma Berenices and includes what we know as the Virgo cluster of galaxies (Fig. 3.8). Herschel wonders if it might perhaps “even make a circuit of the heavens”, although he is uncomfortable about suggesting it intersects our Galaxy exactly where the Sun is situated (p. 111). Herschel, pioneering the study of “the interior construction of the heavens”, has not given himself time to think these matters through.

In its final lines the paper returns to its primary purpose, the establishment of a case for a monster reflector:

> . . . so much we may venture to hope, that, by applying ourselves with all our powers to the improvement of telescopes, which I look upon as yet in their infant state, and turning them with assiduity to the study of the heavens, we shall in time obtain some faint knowledge of, and perhaps be able partly to delineate, *the Interior Construction of the Universe* (p. 112).

4

The riddle of the nebulae: (2) Nebulosity rejected

When William Herschel sent to the Royal Society his 1784 paper on the construction of the heavens, he was still of the opinion that there were true nebulae to be found alongside those that were simply distant "clusters of stars in disguise", and he had persuaded himself that observers could tell the difference: true nebulae shone with a milky appearance, while star clusters disguised by distance appeared mottled, 'resolvable'.

As is the way of things, just five days after the paper was read to the Royal Society, Herschel came across M17, which we know as the Omega Nebula. To his surprise it contained both nebulosities, milky and 'resolvable'. Already convinced that the Milky Way is the optical effect of our immersion in a layer or 'stratum' of stars (the Galaxy), and that he had encountered the remnants of other strata that had already fragmented into nebulae and clusters as a result of gravitational attraction, he imagined the nebula as a stratum of stars, the stars in the middle distance appearing to him as resolvable nebulosity and those in the far distance appearing as milky. But this would require a dramatic reversal of the theory he had just shared with the Royal Society.

> A wonderful Nebula. . . . It is not of equal brightness throughout, and has one or more places, where the milky nebulosity seems to degenerate into the resolvable kind. Should this be confirmed on a very fine night, it would bring on the step between these two nebulosities which is at present wanting, and would lead us to surmize that this nebula is a stupendous Stratum of immensely distant fixed stars some of whose branches are near enough to us to be visible as resolvable nebulosity, while the rest runs on to so great a distance as only to appear under the milky form.[1]

Four weeks later, he encountered M27, the Dumbbell Nebula, and again he found both nebulosities in a single nebula. It was, he decided, another stratum of stars, and indeed its edge nearest to Earth was so close that he could even make out some of the individual stars.

> I suppose it to be a double stratum of stars of a very great extent. The ends next to us are not only resolvable nebulosity but I really do see very many of the stars mixt with the resolvable nebulosity. Farther on the nebulosity is but rarely resolvable & ends at last in milky whitishness of the same appearance as that in Orion.[2]

So true nebulosity was a figment of the imagination, nebulae were simply star clusters after all, and Herschel lived in a universe that was populated by stars accompanied by their planets and comets. The changes he had himself observed in the Orion Nebula must be illusory.

With the long-standing puzzle concerning nebulosity apparently solved, Herschel prepared a second paper on the construction of the heavens, dated New Year's Day 1785 and one of the greatest in the history of astronomy (below, pp. 113–135).[3] He begins in dramatic fashion with an imagined model of a stellar universe in its infancy: "Let us then suppose numberless stars of various sizes, scattered over an indefinite portion of space in such a manner as to be almost equally distributed throughout the whole" (p. 114). Gravity is the agent that will bring about change in this distribution: "The laws of attraction, which no doubt extend to the remotest regions of the fixed stars, will operate in such a manner as most probably to produce the following remarkable effects." And he goes on to describe the "Formation of Nebulae" – that is, star systems – likely to result from irregularities in the hypothetical distribution, places where the gravitational pull is greater than normal.

He instances the presence in a particular region of either one large star, or unusually many ordinary ones. The large star will attract those stars immediately around it, and this will result in a regular (spherical, or near-spherical) cluster of stars – Form I. The unusually numerous ordinary stars will similarly draw in those around, and this will result in an irregular cluster, its shape depending on the layout of these ordinary stars – Form II. Combinations of these situations will result in more complex configurations, Forms III and IV, as well as in Form V, "great cavities or vacancies by the retreat of the stars towards the various centers which attract them" (p. 116).

Does this mean, he asks himself, that clusters will one day end in what we would term 'gravitational collapse'? "At first sight then it will seem as if a system, such as it has been displayed in the foregoing paragraphs, would evidently tend to a general destruction, by the shock of one star's falling upon another." As we saw in Chapter 2, a century earlier Isaac Newton too had worried about the role of gravity in the universe – which for him was strictly infinite – and he was puzzled as to why the stars still occupied the positions listed for them in Antiquity (as the evidence then suggested); for forces cause movements, and if each star was being pulled by the gravitational attraction of every other star it should surely respond by moving. Newton's conclusion had been that Providence had established stars throughout the universe with a high degree of symmetry, so that each star would be pulled (more or less) equally in all directions and therefore in the short term would remain at rest. But since the symmetry was not perfect, sooner or later stars would begin to move. However, all was not lost, for God the Clockmaker had a servicing contract with his universe and would intervene at intervals, and by Divine fiat restore the original distribution of the starry universe.[4]

Unknowingly following in Newton's footsteps, Herschel indulges in some reflections concerning his own "theoretical view". He comments that in large measure "the indefinite extent of the sidereal heavens . . . must produce a balance that will effectually secure all the great parts of the whole from approaching to each other" (p. 116). And, again like Newton, "there is no doubt that the great Author of [the system of the universe] has amply provided for the preservation of the whole, though it should not appear to us in what manner this is effected" (p. 116). Herschel goes on to propose a third consideration, one not available to Newton: if the stars of a cluster are in rotation, then the projectile forces "will prove such a barrier against the seeming destructive power of attraction as to secure from it all the stars belonging to a cluster, if not for ever, at least for millions of ages" (p. 116).

But what then?

> Besides, we ought perhaps to look upon such clusters, and the destruction of now and then a star, in some thousands of ages, as perhaps the very means by which the whole is preserved. These clusters may be the *Laboratories* of the universe, if I may so express myself, wherein the most salutary remedies for the decay of the whole are prepared (p. 116).

Planetary nebulae

He explains what he has in mind in the concluding section of the paper (pp. 133–135), where he lists half-a-dozen examples of what he calls "planetary nebulae", the term we still use today. Herschel had come across the first of these mysterious objects within weeks of taking up his duties at Windsor. On 7 September 1782 he found near ν Aquarii:

> A curious Nebula, or what else to call it I do not know. it is of a shape somewhat oval, nearly circular, and with this power [460] appears to be 10 or 15″ diameter. It is of the same shape with 278, but much less in appearance. with 932 it is still the same shape but much larger. So that its appearance seems to follow the law of magnifying, from whence it is clear that it is of some real magnitude in the heavens and not a glare of light. The brightness in all the powers does not differ so much as if it were of a planetary nature, but seems to be of the starry kind, tho' no star is visible with any power. It is all over of the same brightness. The compound eyepiece will not distinguish it from a fixt star, at least not sensibly. . . .[5]

This momentous discovery of what we know as the Saturn Nebula was to have far-reaching consequences for Herschel's theorizing. The object appeared to have the disk of a planet but the pale light of a nebula; but whether it was a true nebula he was by no means sure. He returned to the object repeatedly to check whether it had changed its position (if so, it would be a planet of sorts);[6] he vacillated, but finally concluded it had not.

In his 1785 paper Herschel carefully considers the various possible explanations of these objects "that from their singular appearance leave me almost in doubt where to class them", and concludes that they are most likely nebulae that "consist of stars that are compressed and accumulated in the highest degree" – in other words, the terminal stage of a globular ('spherical') star system that has been condensing more and more as gravity works its magic.

> If it were not perhaps too hazardous to pursue a former surmise of a renewal in what I figuratively called the Laboratories of the universe, the stars forming these extraordinary nebulae, by some decay or waste of nature, being no longer fit for their former purposes, and having their projectile forces, if any such they had, retarded in each other's atmosphere, may rush at last together, and either in succession, or by one general tremendous shock, unite into a new body (p. 135).

The star that flared up in 1572 might, he thinks, have been an example of such a gravitational collapse.

This was Herschel's best guess as to the nature of planetary nebulae, but they continued to worry him. Soon after the paper was published we find him writing to the French astronomer Lalande describing planetaries as "celestial bodies of which as yet we have no clear idea and which are perhaps of a type quite different from those that we are familiar with in the heavens".[7] And astronomers who visited Slough would be shown a planetary nebula and asked what they thought of it.[8]

The Galaxy

The readers of Herschel's 1785 paper must have been puzzled as to what to make of his near-uniform, indefinite (does he mean 'infinite'?) distribution of stars, for his commitment to the 'stratum' that is our Galaxy was unwavering (and much of the rest of the paper is devoted to an analysis of it). But surely there was no way in which gravity could derive strata from a near-uniform *infinite* distribution of stars. We probably have to understand Herschel as, for the sake of simplicity, talking about a purely hypothetical distribution of stars over "an indefinite portion of space", although in fact for him the local reality is a distribution that can be thought of as sandwiched between two parallel (albeit widely separated) planes.[9]

The observations he has so far had time to make lead him to think that the Galaxy is "every where terminated" (p. 119), finite. But he hesitates to be dogmatic on this. Persevering in his campaign to persuade the King to fund a large reflector, he comments:

> I ought also to add, that a telescope with a much larger aperture than my present one, grasping together a greater quantity of light, and thereby enabling us to see farther into space, will be the surest means of compleating and establishing the arguments that have been used: for if our nebula is not absolutely a detached one, I am firmly persuaded, that an instrument may be made large enough to discover the places where the stars continue onwards (pp. 121–122).

Prophetic words

Not only is the Galaxy seemingly finite, but it has "a certain air of youth and vigour" (p. 127), and it displays "fewer marks of profound antiquity" (p. 127) than other systems; even though "Some parts of our system indeed seem already to have sustained greater ravages of time than others" (p. 128), gravity has not yet disturbed the primeval uniformity of distribution of the stars of our Galaxy sufficiently to invalidate the investigation in which Herschel has been involved for some time.

In his 1784 paper Herschel had introduced his readers to 'star-gages', the first example in history of stellar statistics. He was occupied, he told them, in counting the number of stars in his field of view in various directions, "to come to a full knowledge of the sun's place in the sidereal stratum" (p. 108). He develops the method in his 1785 paper. Granted that his 20-ft reflector can reach the borders of the Galaxy in every direction (as he thinks to be the case), and if within these

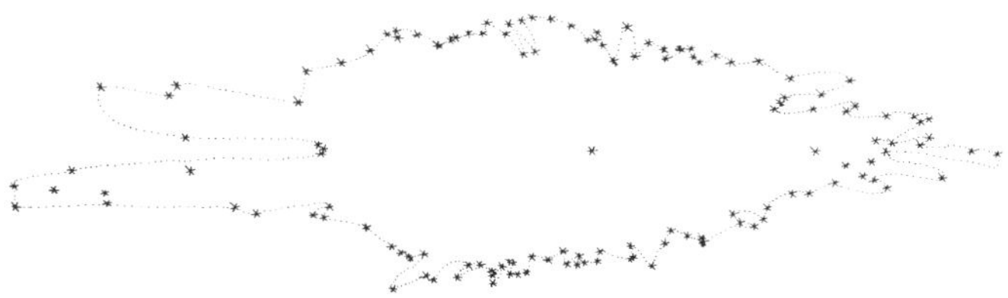

Fig. 4.1 Herschel's preliminary sketch for the cross-section of the Galaxy he published *Philosophical Transactions* in 1785 (p. 126). RAS W.4/24.1, courtesy of the Royal Astronomical Society.

borders the stars are still distributed with fair uniformity, then the number of stars in any field of view is related (by a simple arithmetical formula) to the distance to the border of the Galaxy in that direction. He has been, he says, unable to spare the time to count stars in *every* region of the sky visible from Windsor, a mammoth undertaking, but he has been able to count them around a great circle of the heavens. He has then used the resulting 'star-gages' to derive a diagram representing a cross-section of the Galaxy (p. 126; for a preliminary sketch see Fig. 4.1). So famous would the diagram become that it was being reproduced long after his death, and longer still after he abandoned the two assumptions on which the diagram was based; for writers on astronomy, like Nature, abhor a vacuum, and when the diagram was discredited there was nothing equivalent to take its place.

But discredited it was. After Herschel commissioned his giant 40-ft reflector (Fig. 4.2) in 1789, he found he was able to see many galactic stars invisible in the 20-ft, so the first assumption was incorrect, and there was no reason to suppose that the 40-ft had succeeded where the 20-ft had failed.[10] As to the assumption of near-uniformity, long nights of searching for star clusters would (as we shall see) eventually convince him that a high star-count was an indication of clustering, rather than of an exceptional distance to the border of the Galaxy in that direction.

But this was for the future. For the present it seemed that the compound nebula (p. 125) we know as the Galaxy was finite, indeed compact. The other nebulae (he believed) were likewise star systems, of varying extent; and if the Orion Nebula, for example, or the Andromeda Nebula stretched across the sky and yet was so distant that its individual stars could not be detected, it "cannot be otherwise than of a wonderful magnitude, and may well outvie our milky-way in grandeur" (p. 130). In other words, such nebulae were galaxies fully comparable with our own.

Our Galaxy was destined eventually to fragment, as time passed and gravity undermined the initial uniformity of distribution: "our system, after numbers of ages, may very possibly become divided so as to give rise to a stratum of two or

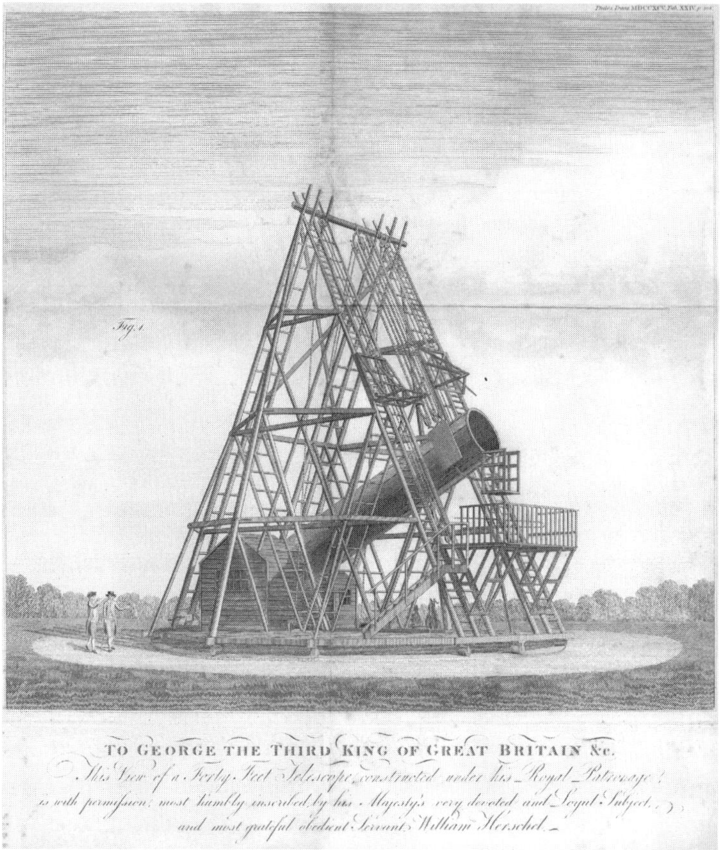

Fig. 4.2 An engraving of Herschel's 40-ft reflector, with its dedication to King George III, who had funded its construction (albeit with growing reluctance). From Herschel's long description of the instrument published in *Philosophical Transactions* in 1795.

three hundred nebulae" (p. 127). The future, then, was clear; but the past was not: as to how the initial near-uniformity had come about, Herschel had no suggestions to offer, and never would have.

"Rise, progress, and decay"

The third of his cosmological papers of the 1780s takes the modest form of "a few introductory remarks on the construction of the heavens" (pp. 136–145) as preface to his second catalogue of one thousand new nebulae and clusters, and is dated 1 May 1789.[11] By now the 40-ft was funded, indeed nearing completion, so there was no longer any need to argue the case for 'more light'; and his own unrivalled familiarity with the large-scale universe was beyond dispute. He begins "*analyzing* the heavens" by making the obvious point that of the bodies in the solar system it is only the Sun, "a refulgent fountain of light", that would be visible from a distance; and so it is only the stars, or "suns, every one of which is probably of as much consequence to a system of planets, satellites, and comets, as our own sun", that the student of the large-scale universe can see and study (pp. 136–137). But he has a dramatic new way of looking at the specimens of 'systems' – star clusters – that he, the natural historian of the heavens, is collecting.

> ... the heavens consist of regions where suns are gathered into separate systems; ... but may we not hope that our knowledge will not stop short at the bare enumeration of phaenomena capable of giving us so much instruction? Why should we be less inquisitive than the natural philosopher, who sometimes, even from an inconsiderable number of specimens of a plant, or an animal, is enabled to present us with a history of its rise, progress, and decay? (p. 137)

Most of the paper is devoted to the many systems he has encountered that have the form of "lucid spots, of equal lustre, scattered over a circular space, in such a manner as to appear gradually more compressed towards the middle" (p. 137). He has no trouble in arguing that these are globular clusters, as indeed they are. They vary in size, in distance, and in the extent to which they are currently condensed, but he sees them all as members of the same species (pp. 139, 143). As a good astronomer Herschel preferred a simple theory to a more complex, and his thinking about celestial objects was invariably guided – for better or worse – by the image of the life-cycle of a *single* living organism with its unique trajectory through time.[12]

The spherical structure of these clusters demonstrates that they "are thus formed by the action of central powers" (p. 140). Gravity is the obvious candidate; Herschel cannot prove that it is the one and only attractive power at work, but in his heart he has no doubt that this is the case, and in later papers he will take this for granted.[13]

A globular cluster (his Form I) will become more and more compressed as time passes, culminating in a planetary nebula. But what of the clusters that are irregular? "I shall now extend the weight of my argument, by taking in likewise every cluster of stars or nebula that shows a gradual condensation, or encreasing brightness, towards a center or certain point" (p. 142). But he has the problem that gravity alone will hardly suffice to make an irregular cluster spherical. However, he is determined to make his theory as simple as possible, and so he is driven to speculate.

> Although the form of [irregular clusters] be not globular, it is plainly to be seen that there is a tendence towards sphericity, by the swell of the

> dimensions the nearer we draw towards the most luminous place, denoting as it were a course, or tide of stars, setting towards a center. And – if allegoral expressions may be allowed – it should seem as if the stars thus flocking towards the seat of power were stemmed by the crowd of those already assembled, and that while some of them are successful in forcing their predecessors sideways out of their places, others are themselves obliged to take up lateral situations, while all of them seem equally to strive for a place in the central swelling, and generating spherical figure (p. 142).

Even his best friend found this unconvincing. Within days of the paper's being read to the Royal Society, William Watson wrote to him with detailed criticisms, among them:

> May not other collections of Stars have been equally exposed to the central powers as the round ones, & yet not arrived to that figure from the previous more unfavorable position of its stars? – If not, show the improbability of the latter hypothesis.

In his covering letter Watson emphasizes the point:

> For tho' I may not be able to produce any reason why irregular clusters have not become round; it should seem you ought to shew that the previous [*or* primaeval] arrangement in such clusters was similar to that of the globular ones & that therefore time only was required for the effect of making them globular likewise.[14]

We might think that Watson's objections were well founded. But Herschel was rarely persuaded into changing his mind, and although some of the clusters he encountered in the heavens were globular and others irregular, his instinct was against accepting that they formed two different species (as they in fact do). To him, even an irregular cluster would eventually develop into a globular; then it would be "in the perfection of its growth", eventually to condense to become a tightly packed planetary nebula, "very aged, and drawing on towards a period of change, or dissolution" (p. 144).

Herschel's concluding paragraph in the 1789 paper epitomizes the revolution in astronomical thinking that he is pioneering:

> This method of viewing the heavens seems to throw them into a new kind of light. They now are seen to resemble a luxuriant garden, which contains the greatest variety of productions, in different flourishing beds; and one advantage we may at least reap from it is, that we can, as it were, extend the range of our experience to an immense duration. For,

to continue the simile I have borrowed from the vegetable kingdom, is it not almost the same thing, whether we live successively to witness the germination, blooming, foliage, fecundity, fading, withering, and corruption of a plant, or whether a vast number of specimens, selected from every stage through which a plant passes in the course of its existence, be brought at once to our view? (p. 145).

5

The riddle of the nebulae: (3) Nebulosity confirmed

A sighting of "luminous fluid"

With the publication of Herschel's 1789 "Remarks", it seemed as though the essentials of his theory of the construction of the heavens were established. But there were problems. One concerned the Orion Nebula. On the supposition that it was a star system so far away that not even he could resolve it into its component stars, Herschel in 1785 had declared that it "may well outvie our milky-way in grandeur" (below, p. 130), that it was a galaxy. But this was to close his eyes to the changes in the nebula that he believed he had observed in his early years as an astronomer. As he was eventually to write, in 1802, "The changes I have observed in the great milky nebulosity of Orion, 23 years ago, and which have also been noticed by other astronomers, cannot permit us to look upon this phaenomenon as arising from immensely distant regions of fixed stars" (p. 166).

Then there were the planetary nebulae, "heavenly bodies, that from their singular appearance leave me almost in doubt where to class them" (p. 133). His best guess was that they were compressed globular clusters on the verge of gravitational collapse (p. 135); but what happened then?

A planetary nebula is in fact a cloud of gas that has been ejected from a dying star, but in the examples Herschel had so far encountered, the star – if visible at all in his reflectors – had not been bright enough to force itself on his attention. This was about to change. At 3.51 a.m. on 13 November 1790, he was at the eyepiece of the 20-ft engaged in a routine sweep, number 980 in the series, with Caroline at her desk at a nearby window. There was no reason to expect that something dramatic was about to happen. But, as the sky slowly rotated overhead, a bizarre and ghostly object came into his field of view: a star with an

Fig. 5.1 The planetary nebula we know as NGC 1514, which Herschel came across on 13 November 1790 and which he interpreted as a 'nebulous star' condensing out of the surrounding nebulosity. It was this observation that convinced him of the existence of 'true nebulosity'. Image courtesy of Dietmar Hagar, FRAS, © www.stargazer-observatory.com.

atmosphere (Fig. 5.1). Caroline copied down the description at her brother's shouted dictation:

> A most singular phenomenon! A star about 8m, with a faint luminous atmosphere of a circular form, of about 3′ dia[mete]r. The star is perfectly in the center, and the atmosphere is so diluted, faint, and equal throughout, that there can be no surmise of its consisting of stars; nor can there be a doubt of the evident connection between the atmosphere and the star.[1]

Here, surely, was an indisputable example of true nebulosity.

As he went back over his records, Herschel could scarcely believe the number of occasions when – with a mind closed to the possibility – he had come across other examples of what he now recognized to be nebulosity. For example, on 16 October 1784 he had encountered "A star of about the 9th magnitude, surrounded by a milky nebulosity, or chevelure, of about 3 minutes in diameter" (NGC 2170). As he now admitted in the paper he wrote for *Philosophical Transactions* (pp. 146–156; see p. 151),

> My present judgement concerning this remarkable object is, that the nebulosity belongs to the star which is situated in its center. . . . It must

> appear singular, that such an object should not have immediately suggested all the remarks contained in this Paper...[2]

and he is hard put to it to find excuses.

Hitherto Herschel had seen planetary nebulae as representing the final stage in the evolution of stellar systems, globular clusters about to collapse and thereby become one of the “laboratories of the universe” (p. 116). Now, in a dramatic change of heart, he would see a planetary as pre-stellar, as “a much condensed, luminous fluid” (p. 155), soon to develop under gravity into a nebulous star, and thence into a true star.[3] As to the “atmosphere” surrounding the star he encountered in November 1790, “If, therefore, this matter is self-luminous, it seems more fit to produce a star by its condensation than to depend on the star for its existence” (p. 154).

It is ironic that while to modern astronomers the star with atmosphere is a planetary nebula (NGC 1514), to Herschel it was not a planetary nebula at all but a ‘nebulous star’ – by definition, Herschel’s planetary nebulae had uniform brightness, and this object was far from uniform.

Herschel hints at the possible physical origins of the luminous fluid:

> How far the light that is perpetually emitted from millions of suns may be concerned in this shining fluid, it might be presumptuous to determine; but, notwithstanding the unconceivable subtilty of the particles of light, when the number of emitting bodies is almost infinitely great, and the time of continual emission indefinitely long, the quantity of emitted particles may well become adequate to the constitution of a shining fluid, or luminous matter, provided a cause can be found that may retain them from flying off, or reunite them (pp. 155–156).

The individual particles of light will be travelling at immense speed in all directions, but as they “pass through innumerable systems” they will meet obstacles that will impede their flight, “Not to mention the great counteraction of the united attractive force of whole sidereal systems, which must be continually exerting their power upon the particles while they are endeavouring to fly off” (p. 156). As a result, he seems to think, they will slow down, and so be able to attract surrounding particles and form luminous fluid.

The final cosmogony

Herschel’s recognition of true nebulosity called for a major rethink of his cosmogony. Whereas his 1785 theoretical model of the universe had begun

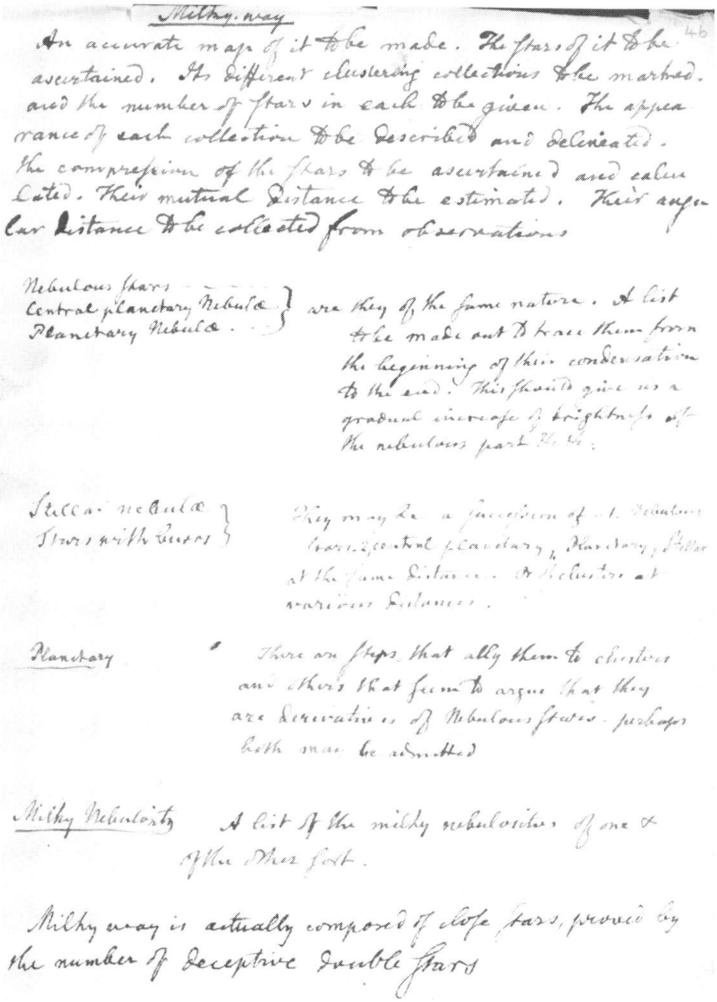

Milky-way

An accurate map of it to be made. The stars of it to be ascertained. Its different clustering collections to be marked. and the number of stars in each to be given. The appearance of each collection to be described and delineated. The compression of the stars to be ascertained and calculated. Their mutual distance to be estimated. Their angular distance to be collected from observations

Nebulous stars
Central planetary Nebulæ
Planetary Nebulæ } are they of the same nature. A list to be made out to trace them from the beginning of their condensation to the end. This should give us a gradual increase of brightness of the nebulous part &c &c.

Stellar nebulæ
Stars with burs } They may be a succession of 1. Nebulous stars, 2. central planetary, Planetary, Stellar at the same distance. Or clusters at various distances.

Planetary — There are steps that ally them to clusters and others that seem to argue that they are derivatives of Nebulous stars. perhaps both may be admitted

Milky Nebulosity — A list of the milky nebulosities of one & of the other sort.

Milky way is actually composed of close stars, proved by the number of deceptive double stars

Fig. 5.2 Most of Herschel's rough drafts were later destroyed by his family, to the frustration of historians, but this early outline for his 1802 "Remarks on the construction of the heavens" survived. RAS W.7/4, f. 46, courtesy of the Royal Astronomical Society.

with a nearly uniform distribution of fully formed stars on which gravity went to work, his new picture of the real universe envisaged pre-stellar stages during which the clouds of nebulosity condensed into nebulae. These eventually gave birth to stars, which fed on any remaining nebulosity and then went on (as before) to develop into clusters. This revised cosmogony, which he was to accept for the rest of his days, he first expounded in the "Remarks on the construction of the heavens" prefaced to his final catalogue of nebulae, published in 1802 (pp. 157–168).[4]

In 1784 (p. 107) Herschel had invoked what we today term the Copernican Principle, that the Earth's position in the universe is in no way special, to justify his belief that the Sun is a typical star of the Galaxy. But in 1802, at the very time that Caroline was assembling the nebulae for their final catalogue, he was in process of re-examining his double stars after an interval of some two decades, and he had found that some were in fact not line-of-sight doubles occurring by chance, but 'binaries', formed of two stars that were companions in space and orbiting together about their common centre of gravity (p. 163). Were either of the two stars of a binary to have had in the past its own planetary system, the gravitational pull of the other star would long since have disrupted the orbits of these planets. Herschel concluded that only truly isolated stars – such as the Sun and the bright stars near to us – had the untroubled gravitational conditions that allowed for planets, and that the innumerable stars in multiple systems or clusters would therefore be unable to have satellites in orbit about them (p. 159).

In 1785, early in his long campaign of sweeps for nebulae and clusters, Herschel had seen the stars within the Galaxy as distributed with fair uniformity (p. 120), but by 1802, at the end of sweeps during which he had discovered many hundreds of clusters, he had rejected this view: the stars of the Galaxy, he says, are in fact "very unequally scattered" (p. 163). Perhaps his diagram of a cross-section of the Galaxy was becoming an embarrassment, for in later life he repeatedly disowns the assumption of equal scattering on which it was based.[5] To the contrary,

> . . . the stars we consider as insulated [including the Sun] are also surrounded by a magnificent collection of innumerable stars, called the milky-way, which must occasion a very powerful balance of opposite attractions, to hold the intermediate stars in a state of rest. For, though our sun, and all the stars we see, may truly be said to be in the plane of the milky-way, yet I am now convinced, by a long inspection and continued examination of it, that the milky-way itself consists of stars very differently scattered from those which are immediately about us. But of this, more will be said on another occasion (p. 159).

This sounds almost like a ring theory of the Galaxy, but this is not his meaning, for as he was to explain in 1818, the Galaxy extends without visible limits: ". . . when our gages will no longer resolve the milky way into stars, it is not because its nature is ambiguous, but because it is fathomless". This had implications. The Andromeda Nebula, for example, may possibly be another stellar system, but this is by no means certain, for "its nature remains mysterious" (p. 194). If it is a stellar system, it is evidently of finite extent. By contrast our Galaxy, being fathomless in its length and breadth, is unique, "the most brilliant, and beyond all comparison the most extensive sidereal system";[6] and so again the Copernican Principle does not apply.

In the 1802 "Remarks" he goes on to discuss "groups of stars", formerly referred to as "irregular clusters" and "perhaps, of all the objects in the heavens, the most difficult to explain" (p. 164). However, for once he is agnostic about what the future holds for them and makes no attempt to argue that they develop under gravity into globular clusters (which he esteems as "certainly the most magnificent objects that can be seen in the heavens" (p. 164)).

The paper ends (pp. 166–167) with an account of milky nebulosity, nebulous stars, planetary nebulae, and planetary nebulae with centres. Milky nebulosity, he says, may sometimes simply be a star system disguised by distance, "like the collections that construct our milky-way", or it may be "real, and possibly at no great distance from us". In the latter case, "To attempt even a guess at what this light may be, would be presumptuous". We know (he argues) that the Orion Nebula is formed of real nebulosity because it has been observed to change, although these changes are in fact "minute". As to the nebulous stars, their nature "is enveloped in much obscurity", but Herschel's best guess is that planetary nebulae develop into planetary nebulae with centres, and these in turn into nebulous stars.

And there the matter rested for nearly a decade, during which Herschel – to the dismay of his friends – became obsessed with explaining the colour effects we know as Newton's rings. But when he was in his seventies, and had long since given up serious observing, Herschel prepared two lengthy papers in which he assembled specimens from his great catalogues of nebulae and clusters, to allow us in effect to 'observe' the life-cycle of these objects to which he had devoted so much effort. In the first, published in 1811 (pp. 169–186), he illustrates the gradual development under gravity of nebulosity and nebulae; in the second, published in 1814 (pp. 187–200), he does the same for nebulae intermingled with stars, and then for the purely "sidereal part of the heavens".[7] There is, he says, less difference between the objects grouped in the successive 'articles' than "there would be in an annual description of the human figure, were it given from the birth of a child till he comes to be a man in his prime" (p. 170).

He believes he has been able to show in the 1811 paper that "every succeeding state of the nebulous matter is the result of the action of gravitation upon it while in a foregoing one" (p. 184). The 1814 paper demonstrates what then follows, the "very gradual conversion of the nebulous matter into the sidereal appearance" (p. 187) and "the *growth* of stars" (p. 190). It then goes on to show how star systems develop under gravity, ultimately to become "the most magnificently constructed sidereal systems" (p. 198), the globular clusters in which "the exertion of a clustering power has brought the accumulation and artificial construction of these wonderful celestial objects to the highest degree of mysterious perfection" (§16). The 1811 and 1814 papers together, he says, reveal

> the intimate connection between the two opposite extremes, one of which is the immensity of the widely diffused and seemingly chaotic nebulous matter, and the other, the highly complicated and most artificially constructed globular clusters of compressed stars (p. 187).

Problems that remained

(i) *The universe in the large*. The limitations of the 1811 and 1814 papers are significant. In them, Herschel deals with the life-stories only of individual astronomical objects – clouds of nebulosity, stars, star clusters – or of groups of closely related objects; and while he has a few remarks to make about the Galaxy, he has nothing to say about the cosmos as a whole. Given Herschel's astonishing boldness and inventiveness as a theorist, this is surprising.

(ii) *The origins and nature of nebulosity*. The 1811 paper begins with examples of "extensive diffused Nebulosity" (p. 171), but as to how this nebulosity originated he is silent (the paper, so to speak, begins with "the birth of a child" but does not explain its conception). The only clue we have been given was in the 1791 paper that first announced the existence of true nebulosity, where he spoke of "the light that is perpetually emitted from millions of suns" (p. 155) and suggested ways in which the light particles might be slowed down to the point where they would attract other particles and coalesce to form nebulosity. In 1802 he was agnostic (p. 166). In 1811 he does no more than hint that we may learn something from the "small telescopic comets as often visit our neighbourhood" (p. 181), for these may be composed of nebulosity, if indeed they are not condensed nebulae; otherwise he is again agnostic: "By nebulous matter I mean to denote that substance, or rather those substances that give out light, whatsoever may be their nature, or of whatsoever different powers they may be possessed" (§2).

Nebulosity, whatever its nature may be, gives out only a pale light, and therefore the only nebulosity within sight is that close to us. But what little we can see is surprisingly extensive, and therefore the total quantity of nebulosity in the universe must be immense. "A nebulous matter", he says in the 1811 paper, "diffused in such exuberance throughout the regions of space, must surely draw our attention to the purpose for which it may probably exist" (§3), and later in the paper he confirms that this purpose is to provide the material that will in time develop into nebulae and thence into stars:

> . . . the present state of the heavens presents us with several extensive collections of scattered nebulae, plainly indicating by their very remarkable arrangement, that they owe their origin to some former stock of nebulous matter (p. 177).

(iii) *The fate of globular clusters.* The 1814 paper ends with globular clusters ("a man in his prime"), but as to what then happens to a cluster as it ages and declines past its current perfection he has nothing to say: he makes no attempt to shed light on the celestial counterpart to old age and death. Back in 1785, when he thought all nebulae were star clusters, he had imagined gravity as bringing about the development of a globular cluster into an even more compact planetary nebula, the final stage before gravitational collapse (p. 135). What does he now think happens to globular clusters? In these later writings there is no suggestion that they are eternal, and surely gravitational collapse is the fate that must await them, even if orbital motions postpone the evil day. One might have expected Herschel to argue that the explosion of such gravitational collapses will result in light flung out into space in all directions, to form nebulosity and so begin the cycle over again. If this was his opinion – as seems very possible – he does not say.

(iv) *How do irregular clusters become globular?* Herschel simply assumes that celestial objects that have been subject to the action of gravity for long enough become spherical. If a cluster is presently irregular in shape, then give it time and eventually it will become a globular cluster: ". . . the still remaining irregularity of their arrangement additionally proves that the action of the clustering power has not been exerted long enough to produce a more artificial construction" (§11).

(v) *What is the origin of the spinning motion of stars?* Here Herschel does have some hints to offer, for he sees the nebula out of which the star will be born as already rotating as a result of irregularities in the nebulosity that formed the nebula, notably the opposite appendages to its nucleus, or 'branches':

> Will not the matter of these branches in their gradual fall towards the nucleus, when discharging their substance into the chevelure, produce a kind of vortex or rotatory motion? . . . do we not see some natural

> cause which may give a rotatory motion to a celestial body in its very formation? (p. 183).

He later enlarges on this:

> . . . it appears that every figure which is not already globular must have eccentric nebulous matter, which in its endeavour to come to the center, will either dislodge some of the nebulosity which is already deposited, or slide upon it sideways, and in both cases produce a circular motion; so that we can hardly suppose a possibility of the production of a globular form without a consequent revolution of the nebulous matter, which in the end may settle in a regular rotation about a fixed axis (§30).

Time past and time future

In his private notes on the geological formation of the Earth itself, Herschel had no hesitation in envisaging "thousands of centuries";[8] and he was still bolder in his thinking about the universe. To look at distant objects was, he said, to look into time past, and he believed he had seen light that had taken two million years on its journey to Earth (p. 166);[9] even if its source had ceased to exist two million years ago, it would still be visible to him. In the development of nebulae, "millions of years, perhaps are but moments" (p. 179). Once he even remarks that "we have an eternity of past duration to resort to" (p. 176), although it is not clear whether he intends 'eternity' in the literal sense or as shorthand for "immense periods of time".

He ends his 1814 paper with some thoughts about the Galaxy. As gravity continues its work within the Galaxy, star clusters will be created which will eventually reach "the ripening period of the globular form", so that the Galaxy "must finally be broken up, and cease to be a stratum of scattered stars" (p. 199). And he concludes with a remark that is at once enigmatic and profound:

> We may also draw a very important additional conclusion from the gradual dissolution of the milky way; for the state into which the incessant action of the clustering power has brought it at present, is a kind of chronometer that may be used to measure the time of its past and future existence; and although we do not know the rate of going of this mysterious chronometer, it is nevertheless certain, that since the breaking up of the parts of the milky way affords a proof that it cannot last for ever, it equally bears witness that its past duration cannot be admitted to be infinite (pp. 199–200).

6

Retrospect: William Herschel and the construction of the heavens

William Herschel's investigations into 'the construction of the heavens' were focused on two questions: Are all nebulae simply clusters of stars disguised by distance?, and How can we explain the appearance of the Milky Way? And his answers depended on two pivotal concepts: in space, that stars are arranged in strata; and in time, that gravity is the agent of change.

He displays a mature grasp of the problem of the nature of the nebulae already in 1774, on the very first page of his first observing book: nebulae that change shape cannot be distant star systems but must be formed of a luminous fluid, true nebulosity. He believed that he had detected changes in the Orion Nebula while still in Bath, and this was still his position in 1784 when he wrote his first paper on the construction of the heavens.

In this paper he gave an enduring answer to the puzzle of the Milky Way: the Sun is immersed in a stratum of stars and this is why we see a milky effect as we look around us within the Galaxy. In 1784 and 1785 he believed that the stratum was of finite extent and that the stars within it were distributed with fair uniformity. This allowed him to use stellar statistics ('star-gages') to plot the outline of a cross-section. But when he commissioned the 40-ft reflector in 1789 he found that the Galaxy extended further than he had imagined – indeed, as far as his instruments could tell, it was without limits. In his 1802 "Remarks" he focused on the contrast between the isolated stars of which the Sun is one and the clusters found elsewhere in the Galaxy, but his concept of the Galaxy as a stratum of stars never wavered. But how such a stratum might have originated was a mystery to which Herschel could offer no solution.

When he wrote his 1784 paper he was already familiar with regions where many nebulae and clusters were to be found. These collections he interpreted as strata analogous to the Galaxy. But it was only after he decided in the summer of

1784 that true nebulosity was an illusion and all nebulae simply clusters of stars, that his cosmos acquired the simplicity that forced gravity – and its consequences – on his attention. In his 1785 paper the strata now consisting of individual clusters were Galaxy-like but were older than the Galaxy, for they had already fragmented as a result of the destructive effects of gravitational attraction.

Gravity is a force, and forces cause movements, changes. His 1785 paper is shot through with this insight. Because strata of stars will fragment under gravity, our now-youthful Galaxy may one day become a stratum of two or three hundred isolated clusters. And because the Orion Nebula and the Andromeda Nebula appear large and yet are too distant for us to detect the stars of which they are composed, they are examples of comparable galaxies.

In his 1789 paper Herschel expresses his vision of the globular cluster as the culmination of a star system's development under gravity. He sees all star systems as forming a single species, and this leads him to imagine that even irregular clusters must somehow join the mainstream of development and eventually become globular. But what happens next? He suspects that planetary nebulae provide the clue, for they may be highly compressed globular clusters, about to undergo gravitational collapse.

Herschel's belated acceptance in 1790 of the existence of true nebulosity required his cosmogony to be extended back in time, to the stages before stars have formed. Planetary nebulae were now given an entirely different role: that of the luminous fluid out of which stars condense. In his 1791 paper Herschel hints that the primitive nebulosity out of which nebulae are formed might originate in the light emitted by the stars of the universe; but this is the only occasion on which he addresses the issue.

His final papers on the construction of the heavens re-present his two great insights, but leave them enveloped in mystery. Nebulae, stars and star clusters are successive stages in the life-cycle of a single species as gravity works its effects, and the stages he sets out for us begin with diffuse nebulosity and end with the perfection of globular clusters; but how the nebulosity arose, and what eventually happens to the globular clusters, we are not told. The Galaxy meanwhile is a stratum of nebulae and stars, and the stratum is of unknown origins and – in two dimensions – of indefinite extent. It is unique, for its one-time rivals (such as the Orion Nebula and the Andromeda Nebula) are visibly finite; indeed, they may be nebulous and not stellar at all.

All this is a far cry from the clockwork universe of Newton and Leibniz, but Herschel left too many questions unanswered, and astronomers were uncomfortable with theories supported by evidence available to one man alone. And so when the future Royal Astronomical Society published in 1820 an "Address ... explanatory of their views and objects", potential members read:

Beyond the limits however of our own system, all at present is obscurity. Some vast and general views of the construction of the heavens, and the laws which may regulate the formation and motions of sidereal systems, have, it is true, been struck out; but, like the theories of the earth which have so long occupied the speculations of geologists, they remain to be supported or refuted by the slow accumulation of a mass of facts. . . .[1]

References

Abbreviations

JHA	*Journal for the History of Astronomy*
Phil. Trans.	*Philosophical Transactions*
RAS	Royal Astronomical Society Herschel Archive

1 The making of an astronomer

1 For details of the life of William Herschel and his sister Caroline, see Michael Hoskin, *Discoverers of the Universe: William and Caroline Herschel* (Princeton, 2011).

2 For biographies of Isaac and of each of his children, see Michael Hoskin, *The Herschels of Hanover* (Cambridge, 2007).

3 Private and Personal Acts 1793, *c.* 38.

4 Caroline's story is told in her own words in *Caroline Herschel's Autobiographies*, ed. by Michael Hoskin (Cambridge, 2003), and more generally in Michael Hoskin, *The Herschel Partnership: As Viewed by Caroline* (Cambridge, 2003).

5 Robert Smith, *A Compleat System of Opticks* (2 vols, Cambridge, 1738).

6 James Ferguson, *Astronomy Explained upon Sir Isaac Newton's Principles* (1st edn, London, 1756). Herschel bought a copy of one of the many later editions.

7 The classic article on Herschel's telescopes is J. A. Bennett, "'On the power of penetrating into space': The telescopes of William Herschel", *JHA*, **7** (1976), 75–108. See also Andreas Maurer, "A compendium of all known William Herschel telescopes", *Journal of the Antique Telescope Society*, no. 14 (1998), 4–15.

8 Michael Hoskin, "Vocations in conflict: William Herschel in Bath, 1766–1782", *History of Science*, **41** (2003), 315–333.

9 RAS W.2/1.1, f. 1.

10 Roy Porter, "William Herschel, Bath, and the Philosophical Society", in G. E. Hunt (ed.), *Uranus and the Outer Planets* (Cambridge, 1982), 23–34.

11 *Caroline Herschel's Autobiographies* (ref. 4), 60, 131.

12 Banks to Herschel, 15 March 1782, RAS W.1/13.B.4.

13 J. A. Bennett, "Herschel's scientific apprenticeship and the discovery of Uranus", in Hunt (ed.), *Uranus and the Outer Planets* (ref. 10), 35–53.

2 Our neighbours among the stars

1 "Unde facile colligitur quod Sol distantia ejus a Terra 900000 vel numero rotundo 1000000 vicibus circiter augeretur", Isaac Newton, Cambridge University Library Add MS 3965, f 279v.

2 Robert Hooke, *An Attempt to Prove the Motion of the Earth* (London, 1674). On this and the subsequent work of Bradley, see Michael Hoskin, *Stellar Astronomy: Historical Studies* (Chalfont St Giles, 1982), chap. A2: "Hooke, Bradley and the aberration of light".

3 James Bradley, "An account of a new-discovered motion of the fixed stars", *Phil. Trans.*, **25** (1727–28), 637–661.

4 James Ferguson, *Astronomy Explained upon Sir Isaac Newton's Principles* (1st edn, London, 1756), §356.

5 Robert Smith, *A Compleat System of Opticks* (2 vols, Cambridge, 1738), **2**, 447.

6 Harald Siebert, "The early search for stellar parallax: Galileo, Castelli, and Ramponi", *JHA*, **36** (2005), 251–271.

7 James Bradley, "A letter . . . concerning an apparent motion observed in some of the fixed stars", *Phil. Trans.*, **45** (1748), 1–43, p. 41. The paper was read on 14 February 1747.

8 RAS W.2/1.1, 50–53.

9 According to Caroline, Maskelyne called "when we lived at Walcot in 1777.78", Michael Hoskin, *Caroline Herschel's Autobiographies* (Cambridge, 2003), 60. In fact they left Walcot on 29 September 1777, *ibid.*, 137.

10 RAS W.2/1.1, 54.

11 The usually reliable Dreyer (J. L. E. Dreyer (ed.), *The Scientific Papers of Sir William Herschel* (2 vols, London, 1912), **1**, p. xxix) thinks the review yielded a handful of double stars, but this seems not to be the case. It is likely that the second review (begun 17 August 1779) followed hard on the heels of the first.

12 William Herschel, "Catalogue of double stars", *Phil. Trans.*, **72** (1782), 112–162.

13 After 1 October 1779 and before 3 October.

14 RAS W.1/13.M.18.

15 The Committee was only repeating formally an objection that Herschel was already well aware of, for on 1 February 1782 he wrote to Hornsby seeking to justify the postulate, RAS W.1/1, p. 33.

16 John Michell, "An inquiry into the possible parallax and magnitude of the fixed stars, from the quantity of light which they afford us, and the particular circumstances of their situation", *Phil. Trans.*, **57** (1767), 234–264.

17 Hoskin, *Caroline Herschel's Autobiographies* (ref. 9), 62.

18 Michell, "An inquiry" (ref. 16), 246–249.

19 *Ibid.*, 240–241.

20 *Ibid.*, 241–242.

21 Isaac Newton, Cambridge University Library Add MS 3965, f 279r–280v. For a transcript and discussion see Michael Hoskin, "Newton, Providence and the universe of stars", *Stellar Astronomy* (ref. 2), chap. B2.

22 Edmond Halley, "Of the infinity of the sphere of the fix'd stars", *Phil. Trans.*, **31** (1720–21), 22–24, read to the Royal Society on 9 March 1720/1 Old Style, and "The

number, order, and light of the fix'd stars", *ibid.*, 24–26, read on 16 March 1720/1. Michael Hoskin, "Halley and 'Olbers's Paradox'", *Stellar Astronomy* (ref. 2), 95–100.

23 Christian Mayer, "Verzeichniss aller bisher entdeckten Doppelsterne", *Astronomisches Jahrbuch für das Jahr 1784* (Berlin, 1781), 183–186.

24 Although he did not send the catalogue for publication for another year.

25 John Michell, "On the means of discovering the distance, magnitude etc. of the fixed stars", dated 26 May 1783, *Phil. Trans.*, **74** (1784), 35–57, p. 56.

26 William Herschel, "An account of the changes that have happened, in the relative situations of double stars", *Phil. Trans.*, **93** (1803), 339–382, and "Continuation of an account . . .", **94** (1804), 353–384.

27 Maskelyne to Herschel, 3 March 1803, RAS W.1/13.M.70.

28 Maskelyne to Herschel, 26 May 1803, RAS W.1/13.M.71.

29 Herschel, "An account" (ref. 26), 364.

30 William Herschel, "Catalogue of 500 new nebulae . . . with remarks on the construction of the heavens", *Phil. Trans.*, **92** (1802), 477–528, p. 486.

31 John to Caroline Herschel, 17 April 1832, Constance A. Lubbock, *The Herschel Chronicle* (Cambridge, 1933), 381.

32 For further information on variable stars in the seventeenth century see Hoskin, *Stellar Astronomy* (ref. 2), chap. A2.

33 J. P. Holwarda, *Dissertatio astronomica* (Franeker, 1640).

34 Ismael Bullialdus, *Ad astronomos monita duo* (Paris, 1667), Monitum Primum, De Stella Mirabili quae in Collo Ceti conspicitur (pp. 3–14).

35 Geminiano Montanari, for example, sent the Royal Society an account dated 30 April 1670 of some of his discoveries, adding: "I have observed many more changes among the Fixed Stars, even to the Number of a Hundred", Thomas Birch, *A History of the Royal Society of London* (London, 1756–57), **2**, 448, translation from John Keill, *An Introduction to the True Astronomy* (London, 1721), 58.

36 In a letter of 24 August 1817, Pigott reminds Herschel of "the Astronomical Observations we made together in 1778 in King Street", RAS W.1/13.P.42.

37 For more on the work of Pigott and Goodricke, see Hoskin, *Stellar Astronomy* (ref. 2), chap. A4.

38 John Goodricke, "A series of observations on . . . Algol", *Phil. Trans.*, **73** (1783), 474–482, p. 482.

39 Hoskin, *Stellar Astronomy* (ref. 2), 43.

40 William Herschel, "On the periodical star α Herculis", *Phil. Trans.*, **86** (1796), 452–482, p. 455. *Cf.* William Herschel, "On the nature and construction of the sun and fixed stars", *Phil. Trans.*, **85** (1795), 46–72, p. 68: "The sun turns on its axis. So does the star Algol. So do the stars β Lyrae, δ Cephei, η Antinoi, o Ceti, and many more; most probably all. From what other cause can we so probably account for their periodical changes? Again, our sun has spots on its surface. So has the star Algol; and so have the stars already named; and probably every star in the heavens. On our sun these spots are changeable. So are they on the star o Ceti; as evidently appears from the irregularity of its changeable lustre, which is often broken in upon by accidental changes, while the general period continues unaltered. The

same little deviations have been observed in other periodical stars, and ought to be ascribed to the same cause."

41 Herschel's four Catalogues of the Comparative Brightness of the Stars appeared in *Phil. Trans.*, **86** (1796), **87** (1797) and **89** (1799).

42 Edmond Halley, "Considerations on the change of the latitude of some of the principal fixt stars", *Phil. Trans.*, **30** (1717–19), 736–738.

43 Bradley, *op. cit.* (ref. 7).

44 *Tobias Mayer's Opera inedita*, transl. and ed. by E. G. Forbes (London, 1791), 112.

45 For the list of eighty stars, see *Opera inedita Tobias Mayer*, **1**, ed. by G. C. Lichtenberg (Göttingen, 1775), 80–81.

46 Forbes (ed.), *op. cit.* (ref. 44), 110–112.

47 J. de La Lande, "Mémoire sur les taches du Soleil, et sur sa rotation", *Mémoires de l'Académie Royale des Sciences* for 1776 (Paris, 1779), 457–514, pp. 513–514.

48 Three volumes of the second edition appeared in Paris in 1771, and the fourth (supplementary) volume in 1781. By that time Mayer's *Opera inedita* had appeared and Lalande was able to include a select list of Mayer 'proper motions' in his §2756.

49 For example, in his comments on the Orion Nebula in §31 of his paper, "Astronomical observations relating to the construction of the heavens", *Phil. Trans.*, **101** (1811), 269–336.

50 Ferguson, *Astronomy* (ref. 4), §367.

51 William Herschel, "On the parallax of the fixed stars", *Phil. Trans.*, **72** (1782), 82–111, p. 104.

52 *Ibid.*, p. 98.

53 William Herschel, "On the proper motion of the sun and solar system", *Phil. Trans.*, **73** (1783), 247–283.

54 Maskelyne to Herschel, 15 March 1783, RAS W.1/13.M.22. On the back of the letter Herschel has drawn a sketch to try and verify Maskelyne's assertion as to the effect on β Cygni of a solar apex at λ Herculis.

55 Aubert to Herschel, 7 March 1783, RAS W.1/13.A.11.

56 First published as "Ueber die Fortrückung unsers Sonnen-Systems", *Astronomisches Jahrbuch für das Jahr 1786* (Berlin, 1783), 259–260. Prévost's apex had the same declination as Herschel's, namely 25 degrees.

57 William Herschel, "On the direction and velocity of the motion of the sun, and solar system", *Phil. Trans.*, **95** (1805), 233–256; "On the quantity and velocity of the solar motion", *ibid.*, **96** (1806), 205–237.

58 On 3 March 1803 Maskelyne wrote to Herschel: "I shall soon have it in my power to inform you of the proper motions of three of the stars, which you inquire about, which belong to my catalogue of 36 stars; and could determine those of the others, by settling their present places and comparing them with Dr Bradley's determinations", RAS W.1/13.M.70. In the event Maskelyne compared his own 1766 and 1802 positions, "Proper motions of the stars, inferred from the 1766 and 1802 catalogues of stars", Cambridge University Library Maskelyne archives, RGO 4/17, 164–165.

59 Herschel, "On the direction and velocity of the motion of the sun" (ref. 57), 234–235.
60 Herschel, "On the proper motion of the sun and solar system" (ref. 53), 277. Michell, in his 1767 paper (ref. 16, 250–260), had discussed at length the system of stars to which the Sun might belong.
61 John Brinkley, *Elements of Astronomy*, 2nd edn (London, 1819), 130.
62 Francis Baily, "On the proper motion of the fixed stars", *Memoirs of the Royal Astronomical Society*, **5** (1833), 147–170, p. 155.
63 F. Argelander, "Über die eigene Bewegung des Sonnensystems", *Mémoires Présentés à l'Académie Impérial des Sciences*, **3** (St Petersburg, 1837), 561–605.
64 E. S. Holden, *Sir William Herschel: His Life and Works* (London, 1881), 154.

3 The riddle of the nebulae: (1) True nebulosity?

1 Edmond Halley, "An account of several nebulae or lucid spots like clouds", *Phil. Trans.*, **29** (1714–16), 390–392. On the early history of nebulae prior to Herschel, including much information not easily available elsewhere, and a translation in full of Messier's catalogue, see Kenneth Glyn Jones, *The Search for the Nebulae* (Chalfont St Giles, 1975).
2 John Michell, "An inquiry into the possible parallax and magnitude of the fixed stars, from the quantity of light which they afford us, and the particular circumstances of their situation", *Phil. Trans.*, **57** (1767), 234–264, pp. 251, 261.
3 William Herschel, "Astronomical observations and experiments, selected for the purpose of ascertaining the relative distances of clusters of stars", *Phil. Trans.*, **108** (1818), 429–470, p. 466. "We may also have surmised nebulae to be no other than clusters of stars disguised by their very great distance", William Herschel, "Astronomical observations relating to the construction of the heavens", *Phil. Trans.*, **101** (1811), 269–336, p. 270 (below, p. 170).
4 Robert Smith, *A Compleat System of Opticks* (2 vols, Cambridge, 1738), **2**, 447.
5 *Ibid.*, 447–448. The fourth star, completing the Trapezium, was found by Picard in 1673.
6 James Ferguson, *Astronomy Explained upon Sir Isaac Newton's Principles* (1st edn, London, 1756), §364. Ismael Bullialdus, *Ad astronomos monita duo* (Paris, 1667), Monitum Secundum, De nebulosa in Andromedae cinguli parte Borea ante biennium iterum orta, reported his chance rediscovery of the Andromeda Nebula in 1664, which he thought must therefore have been invisible in recent decades.
7 *Ibid.*, §365.
8 RAS W.2/1.1, 1.
9 Herschel's rough notes are in RAS W.2/1 under the relevant date, and the fair copy in RAS W.4/1.
10 William Herschel, "Description of a forty-feet reflecting telescope", *Phil. Trans.*, **85** (1795), 347–409, p. 348.
11 J. A. Bennett, "'On the power of penetrating into space': The telescopes of William Herschel", *JHA*, **7** (1976), 75–108, p. 81.

12 RAS W.2/1.1, 43.

13 H. C. King, *The History of the Telescope* (London, 1955), chap. 5.

14 See "William Herschel's lunatick friends" by Andrew Lound, *The Speculum: The Journal of the William Herschel Society*, **9**, part 2 (2010), 11–26, p. 13. For Michell's letter to Watson, see RAS W.1/13.M.99.

15 John Smeaton, the great engineer, wrote to Michell after Smeaton had visited Herschel in 1785 when the 40-ft was under construction, and urged him: "From what I have said, instead of looking at the Satellites of Jupiter, and expecting them to form one Lucid point, or round surface, I would wish you to turn your great Telescope to some of the *Nebulae*, for, as the Surface of your Speculum, is more than double any Mr Herschell *yet* has, it is possible you may make some discoveries as yet unlooked for" (copy of letter dated 4 November 1785, RAS MS Radcliffe Hornsby 78). Michell's inactivity was probably the result of the poor quality of his mirror. After Michell's death Herschel bought the mirror (then in pieces) and examined it, concluding: "Its construction is what I cannot approve . . .", see his "Remarks on Mr Michell's telescope", RAS W.7/14.

16 For an account of this episode, see Michael Hoskin, *Discoverers of the Universe: William and Caroline Herschel* (Princeton, 2011), 53–58.

17 "In a former paper I mentioned, that a more powerful instrument was preparing for continuing my reviews of the heavens. [My new reflector of 18-inches aperture is] far inferior to the one I had undertaken to construct when that paper was written . . . ." William Herschel, "Account of some observations tending to investigate the construction of the heavens", *Phil. Trans.*, **74** (1784), 437–451, p. 437, see below, pp. 99–100.

18 RAS W.2/1.4.

19 RAS C.3/1.2, 57.

20 RAS W.4/1.3, 279.

21 RAS W.4/1.4, 375.

22 Hoskin, *Discoverers of the Universe* (ref. 16), chap. 6.

23 Herschel to Patrick Wilson, 8 March 1784, RAS W.1/1, 99–100.

24 William Watson, Jr, to Herschel, 25 August 1784, RAS W.1/13.W.32.

25 William Watson, Jr, to Herschel, 7 December 1781, RAS W.1/13.W.11. Charles Messier, "Catalogue des nébuleuses et des amas d'étoiles", *Connoissance des Temps pour 1783* (Paris, 1780), 225–249, 408.

26 Herschel, "Account of some observations tending to investigate the construction of the heavens" (ref. 17).

27 See ref. 8.

28 RAS W.4/33.1.

29 Michael Hoskin, "Caroline Herschel as observer", *JHA*, **36** (2005), 373–406, pp. 376–378.

30 Charles Messier, "Catalogue des nébuleuses et des amas d'étoiles", *Connoissance des Temps pour 1784* (Paris, 1781), 227–269. Herschel first observed a nebula (M71) present in this catalogue but not in its predecessor on 30 May 1783; the summer months were far from ideal for delicate observations, and the next such nebula (M92) had to wait until 2 August.

31 His examination of Messier objects culminated on the nights of 30 and 31 July and 2 August, when he observed over twenty of them, most with the 20-ft but some with the 7-ft.
32 An exhaustive analysis of the nebulae and clusters discovered by Herschel is to be found in Wolfgang Steinicke, *Observing and Cataloguing Nebulae and Star Clusters: From Herschel to Dreyer's New General Catalogue* (Cambridge, 2010).
33 All these observations are recorded under the relevant dates in RAS W.4/1.
34 On the night in question, 20 September 1783, Herschel observed M74 and of this there is no doubt as he defines its position exactly; he saw it with some difficulty and says nothing about its appearance. The previous object he observed that night was M27, and the object previous to that is described as "All resolved into exceedingly small stars" in the 7-ft; it too is said to be M74, but this cannot be as M74 is a spiral galaxy. Wolfgang Steinicke offers the plausible suggestion that it was M71, which is not far from M27.
35 David W. Dewhirst and Michael Hoskin "The Rosse spirals", *JHA*, **22** (1991), 257–266.
36 RAS W.4/33.1, facing inside cover.
37 The paper (*op. cit.* (ref. 17)) is dated April 1784 and was read on 17 June.
38 Galileo Galilei, *Sidereus nuncius* (Venice, 1610), 18r.
39 "Anaxagoras, Democritus, and their followers say that the Milky Way is the light of certain stars", Aristotle, *Meteorology*, 345a25–30.
40 Michael Hoskin, "The cosmology of Thomas Wright of Durham", *JHA*, i (1970), 44–52; Thomas Wright of Durham, *An Original Theory or New Hypothesis of the Universe* (London, 1750; facsimile edn ed. by Michael Hoskin, London, 1971).
41 Michael Hoskin, *William Herschel and the Construction of the Heavens* (London, 1963), 115–116.

4 The riddle of the nebulae: (2) Nebulosity rejected

1 RAS W.4/1.7, ff. 642–643.
2 RAS W.2/1.9, f. 29r. M27 is a planetary nebula but the central star, which is fainter than mag. 13, was beyond Herschel's reach.
3 William Herschel, "On the construction of the heavens", *Phil. Trans.*, **75** (1785), 213–266.
4 Michael Hoskin, "Newton, Providence and the universe of stars", *JHA*, **8** (1977), 77–101.
5 RAS W.2/1.2, f. 231.
6 For example, in 1783 on 30 July, 25 August, 20 September, 17 and 23 October and 14 November, and in 1784 on 16 and 28 June, 15 July and 12 November. RAS W.4/1.5, W.4/1.7, W.2/3.2, W.2/3.3.
7 RAS W.1/1, 129, letter to Lalande, 17 March 1785: "...des corps celèstes dont nous n'avons pas encore d'idée bien claire & qui sont peut-être d'un genre tout à fait different de ce que nous connoissons dans les cieux."
8 For example, on 27 November 1787, Herschel showed his visitors the planetary nebula H. IV. 18 (NGC 7662, the Blue Snowball Nebula in Andromeda): "Messrs Cassini,

Mechain Le Seure & Carochet saw this nebula, and the moon being absent, it appeared in its usual planetary view; these Gentlemen saw it very well and admired it as a great curiosity. Mr Cassini observed that a very small fixt star nf the nebula appeared not unlike a sattellite to it" (RAS W.2/3.7). Planetary nebulae were also shown to William Watson and a Mr Marsden on 11 March 1788, to Lord Palmerston on 3 August 1788, to Lalande on 5 August 1788, he "having never before seen a planetary one", to Edward Pigott on 6 August 1788, and to the Abbé Ximenes on 13 April 1789, *ibid.*

9 As he had explained at length in the 1784 paper ("Account of some observations tending to investigate the construction of the heavens", *Phil. Trans.*, **74** (1784), 437–451, pp. 443–445, see below pp. 106–109). There is a minor complication. Because we see the Milky Way as bifurcated, our Galaxy, "the stupendous sidereal system we inhabit", consists of an "extensive stratum and its secondary branch", Herschel, "On the construction of the heavens" (ref. 3), 244 (below, p. 119).

10 "By these observations it appears that the utmost stretch of the space-penetrating power of the 20 feet telescope could not fathom the Profundity of the milky way . . . [the 40 feet] would then leave us again in the same uncertainty as the 20 feet telescope", William Herschel, "Astronomical observations and experiments tending to investigate the local arrangement of the celestial bodies in space", *Phil. Trans.*, **107** (1817), 302–331, p. 327.

11 William Herschel, "Catalogue of a second thousand of new nebulae and clusters of stars; with a few introductory remarks on the construction of the heavens", *Phil. Trans.*, **79** (1789), 212–255, see below pp. 136–145.

12 As we shall see, in 1811 and 1814 he would arrange innumerable nebulae and clusters from his catalogues in sequence, to imitate "the annual description of the human figure, were it given from the birth of a child till he comes to be a man in his prime", William Herschel, "Astronomical observations relating to the construction of the heavens . . .", *Phil. Trans.*, **101** (1811), 269–336, p. 271 (below, pp. 169–186, see p. 170). In 1817 Herschel assumes that "one with another the stars are of a certain physical generic size and brightness, still allowing that all such deviations may exist, as generally take place among the individuals belonging to the same species", *op. cit.* (ref. 10), 309.

13 Herschel, "Astronomical observations relating to the construction of the heavens . . ." (ref. 12), 284 (below, pp. 174–175): "Instead of inquiring after the nature of the cause of the condensation of nebulous matter, it would indeed be sufficient for the present purpose to call it merely a condensing principle; but . . . why should we not look up to the universal gravitation of matter as the cause of every condensation, accumulation, compression, and concentration of the nebulous matter?"

14 Watson to Herschel, 12 May 1789, RAS W.1/13.W.55.

5 The riddle of the nebulae: (3) Nebulosity confirmed

1 RAS W.2/3.8, sweep 980.

2 William Herschel, "On nebulous stars, properly so called", *Phil. Trans.*, **81** (1791), 71–88, p. 78. Wolfgang Steinicke (private communication) comments that Herschel

might well have said the same of two 'planetary nebulae' in which the central star is prominent and which he encountered long before the object that provoked his change of mind: NGC 2170 (IV. 19, but not a planetary nebula in the modern sense), which he viewed on 16 October 1784 and which he cites in the 1791 paper, and NGC 2392 (IV. 45), which he viewed on 17 January 1787. *Cf.* Wolfgang Steinicke, *Observing and Cataloguing Nebulae and Clusters: From Herschel to Dreyer's New General Catalogue* (Cambridge, 2010), 40.

3 *Cf.* William Herschel, "Astronomical observations relating to the construction of the heavens . . .", *Phil. Trans.*, **101** (1811), 269–336, p. 318: ". . . perhaps in progress of time these nebulae which are already in such a state of compression, may be still farther condensed so as actually to become stars".

4 William Herschel, "Catalogue of 500 new nebulae . . . with remarks on the construction of the heavens", *Phil. Trans.*, **92** (1802), 477–528.

5 "when we examine the milky way . . . this supposed equality of scattering must be given up" (Herschel, "Astronomical observations relating to the construction of the heavens . . ." (ref. 3), 270); "it is, however, evident that, if ever it consisted of equally scattered stars, it does so no longer" (William Herschel, "Astronomical observations relating to the sidereal part of the heavens", *Phil. Trans.*, **104** (1814), 248–284, p. 282); his gages of the 1780s, he says in 1817, "relate more immediately to the scattering of stars" than to distance (William Herschel, "Astronomical observations and experiments tending to investigate the local arrangement of the celestial bodies in space", *Phil. Trans.*, **107** (1817), 302–331, p. 325).

6 William Herschel, "Astronomical observations and experiments . . .", *Phil. Trans.*, **108** (1818), 429–470, p. 453.

7 Herschel, "Astronomical observations relating to the construction of the heavens . . ." (ref. 3), and "Astronomical observations relating to the sidereal part of the heavens" (ref. 5).

8 RAS W.7/15, 1798 tour.

9 In September 1813 Herschel met the poet Thomas Campbell and told him the same. Campbell wrote to a friend, "I really and unfeignedly felt at the moment as if I had been conversing with a supernatural intelligence" (Constance A. Lubbock, *The Herschel Chronicle* (Cambridge, 1933), 336, citing *Life and Letters of Th. Campbell*, ed. by William Beattie (London, 1849)).

6 Retrospect: William Herschel and the construction of the heavens

1 "Address . . .", *Memoirs of the Astronomical Society of London*, **1** (1822–25), 4.

PART II HERSCHEL'S COSMOLOGICAL PAPERS IN *PHILOSOPHICAL TRANSACTIONS*

1783: "On the proper motion of the Sun and solar system"

Summary: "From my last observations I give an account of a great number of changes among the fixt stars such as stars that are lost – that have changed their magnitude – that have moved from their former place – new ones come to be visible &c. Hence I conclude that it is highly probable that every star is more or less in motion. By analogy I apply this to the Sun as one of the fixed stars. I deliver afterwards a method that will easiest enable us to detect the direction and quantity of the solar motion and apply it then to observations that have already been made, and facts that are known, whence at last I draw the conclusion that our Sun is now actually moving with a velocity greater than that which the earth has in her annual orbit, towards a point in the northern hemisphere not far from λ or ρ Herculis. . . . I ought to mention also that I have used the theory of Attraction in support of my Hypothesis of the motion of the Solar System."

Herschel to Alexander Wilson of Glasgow, 3 March 1783[1]

On the proper Motion of the Sun and Solar System; with an Account of several Changes that have happened among the fixed Stars since the Time of Mr. *Flamstead*[2]

Read March 6, 1783.

THE new lights that modern observations have thrown upon several interesting parts of astronomy begin to lead us now to a subject that cannot but claim the serious attention of every one who wishes to cultivate this noble science. That several of the fixed stars have a proper motion is now already so well confirmed, that it will admit of no further doubt. From the time this was first suspected

[1] [RAS W.1/1, 65–66.] [2] [*Phil. Trans.*, **73** (1783), 247–283.]

by Dr. HALLEY we have had continued observations that shew Arcturus, Sirius, Aldebaran, Procyon, Castor, Rigel, Altair, and many more, to be actually in motion; and considering the shortness of the time we have had observations accurate enough for the purpose, we may rather wonder that we have already been able to find the motions of so many, than that we have not discovered the like alterations in all the rest. Besides, we are well prepared to find numbers of them apparently at rest, as, on account of their immense distance, a change of place cannot be expected to become visible to us till after many ages of careful attention and close observation, though every one of them should have a motion of the same importance with Arcturus. This consideration alone would lead us strongly to suspect, that there is not, in strictness of speaking, one *fixed* star in the heavens; but many other reasons, which I shall presently adduce, will render this so obvious, that there can hardly remain a doubt of the general motion of all the starry systems, and consequently of the solar one among the rest.

I might begin with principles drawn from the theory of attraction, which evidently oppose every idea of absolute rest in any one of the stars, when once it is known that some of them are in motion: for the change that must arise by such motion, in the value of a power which acts inversely as the squares of the distances, must be felt in all the neighbouring stars; and if these be influenced by the motion of the former, they will again affect those that are next to them, and so on till all are in motion. Now as we know several stars, in divers parts of the heavens, do actually change their place, it will follow, that the motion of our solar system is not a mere hypothesis; and what will give additional weight to this consideration is, that we have the greatest reason to suppose most of those very stars, which have been observed to move, to be such as are nearest to us; and, therefore, their influence on our situation would alone prove a powerful argument in favour of the proper motion of the sun, had it actually been originally at rest. But I shall waive every view of this subject which is not chiefly derived from experience. . . .

[*Herschel outlines his reviews of the heavens (above, pp. 11, 16), and then lists changes already observed among the stars.*]

To return to the principal subject of this paper, which is the proper motion of the sun and solar system: does it not seem very natural, that so many changes among the stars, – many increasing their magnitude, while numbers seem gradually to vanish; – several of them strongly suspected to be new-comers, while we are sure that others are lost out of our sight; – the distance of many actually changing, while many more are suspected to have a considerable motion: – I say, does it not seem natural that these observations should cause a strong suspicion that most probably every star in the heaven is more or less in motion? And though we have no reason to think,[3] that the disappearance of some stars, or new appearance of others, nor indeed the frequent changes in the magnitudes of so many of them are owing to their change of distance from us by proper motions, which could not occasion these phenomena without being inconceivably quick; yet we may well suppose, that motion is some way or other concerned in producing these effects. A slow motion, for instance, in an orbit round some large opaque body, where the star, which is lost or diminished in magnitude, might undergo occasional occultations, would account for some of those changes, while others might perhaps be owing to the periodical return of large spots on that side of the surface which is alternately turned towards us by a rotatory motion of the star. The idea also of a body much flattened by a quick rotation, and having a motion similar to the moon’s orbit by a change of the place of its nodes, whereby more of the luminous surface would one time be exposed to us than another, tends to the same end; for we cannot help thinking with Mr. DE LA LANDE (Mem. 1776), that the same force which gave such rotations, would probably also occasion motions of a different kind by a translation of the center.[4] Now, if the proper motion of the stars in general be once admitted, who can refuse to allow that our sun, with all its planets and comets, that is, the solar system, is no less liable to such a general agitation as we find to obtain among all the rest of the celestial bodies?[5]

Admitting this for granted, the greatest difficulty will be how to discern the proper motion of the sun between so many other (and variously compounded) motions of the stars. This is an arduous task indeed, which we must not hope to see accomplished in a little time; but we are not to be discouraged from the attempt. Let us, at all events, endeavour to lay a good foundation for those who are to come after us. I shall therefore now point out the method of detecting the direction and quantity of the supposed proper motion of the sun by a few geometrical deductions, and at the same time shew by an application of them to some known facts, that we have already some reasons to guess which way the solar system is probably tending its course.

[3] [Herschel changed his mind after drafting this paper. His original version, in which he seeks to explain these changes in a manner compatible with the hypothesis that all stars are physically similar, is reproduced on pp. 97–98.]

[4] Relating to the motion of the fixed stars, the Astronomer Royal has an expression in the second page of the explanation and use of the tables published in his *Astronomical Observations*, which seems to favour this idea, where he mentions the “peculiar but small motions, which many, IF NOT ALL OF THEM, have among themselves, which have been called their *proper motions*, the causes and laws of which are hid for the present in almost equal obscurity.”

[5] See Mr. MICHELL’s note, Phil. Trans. vol. LVII. p. 252.

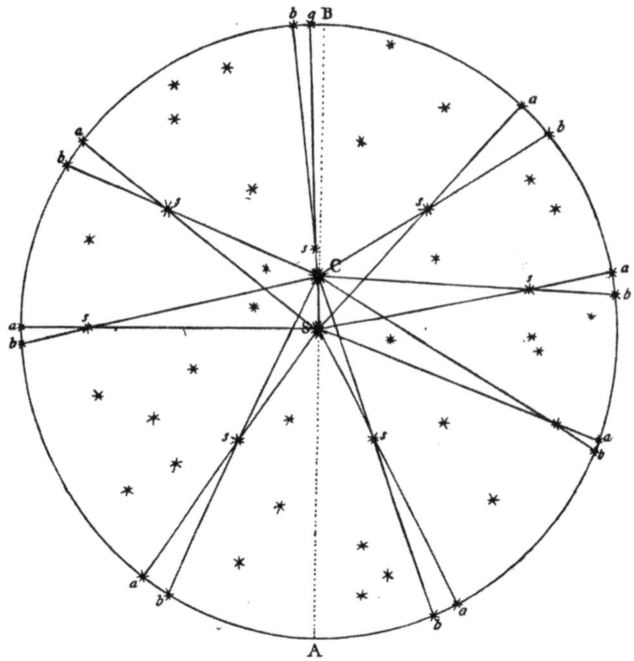

Fig. 1

Suppose the sun to be at S, fig. 1.; the fixed stars to be dispersed in all possible directions and distances around at *s*, *s*, *s*, *s*, &c. Now, setting aside the proper motion of the stars, let us first consider what will be the consequence of a proper motion in the sun; and let it move in a direction from A towards B. Suppose it now arrived at C. Here, by a mere inspection of the figure, it will be evident, that the stars *s*, *s*, *s*, which were before seen at *a*, *a*, *a*, will now, by the motion of the sun from S to C, appear to have gone in a contrary direction, and be seen at *b*, *b*, *b*; that is to say, every star will appear more or less to have receded from the point B, in the order of the letters *ab*, *ab*, *ab*. The converse of this proposition is equally true; for if the stars should all appear to have had a retrograde motion, with respect to the point B, it is plain, on a supposition of their being at rest, the sun must have a direct motion towards the point B, to occasion all these appearances. From a due consideration of what has been said, we may draw the following inferences.

1. The greatest or total systematical parallax of the fixed stars, fig. 2., will fall upon those that are in the line DE, at rectangles to the direction AB of the sun's motion.
2. The partial systematical parallax of every other star, *s*, *s*, *s*, not in the line DE, will be to the total parallax as the sine of the angle BS*a*, being the star's distance from that point towards which the sun moves, to radius.

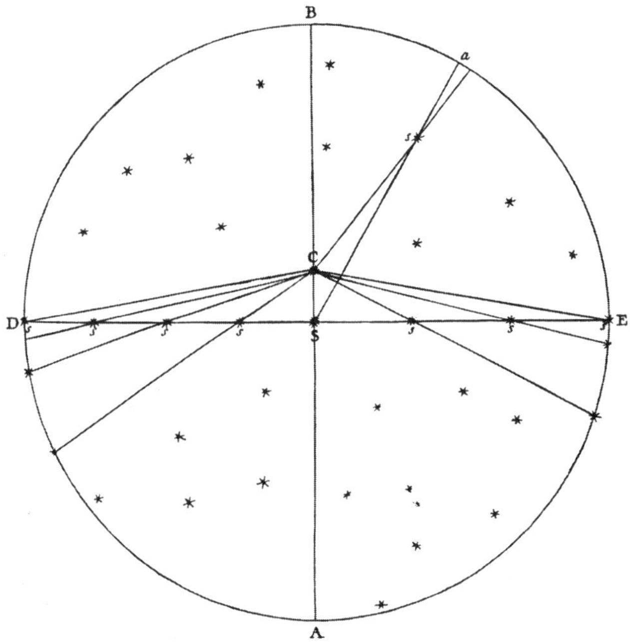

Fig. 2

3. The parallax of stars at different distances will be inversely as those distances; that is, one half at double the distance, one third at three times, and so on; for the subtense SC remaining the same, and the parallactic angle being very small, we may admit the angle S*s*C, to be inversely as the side S*s*, which is the star’s distance.
4. Every star at rest, to a system in motion, will appear to move in a direction contrary to that in which the system is moving.

Corollary. Hence it follows, that if the solar system be carried towards any star situated in the ecliptic: every star, whose angular distance *in antecedentia* (reckoned upon the ecliptic from the star towards which the system moves) is less than 180 degrees, will decrease in longitude. And that, on the contrary, every star, whose distance from the same star (reckoned upon the ecliptic but *in consequentia*) is less than 180 degrees, will increase in longitude, in both cases without alteration of latitude. . . .

[*Herschel lists double stars well placed to reveal the motion of the solar system after a few years of observations, and then returns to his main theme.*]

It remains now only for me to make an application of this theory to some of the facts we are already acquainted with, relating to the proper motion of the stars. And first let me observe, that the rules of philosophizing direct us to refer all phenomena to as few and simple principles as are sufficient to explain them. Thus, for instance, we see the stars and planets rise and set every day: now, as it is much more simple to admit the earth to turn once in 24 hours, than to suppose every single star to revolve round the earth in that time, we very justly ascribe a diurnal motion to the earth; but yet, since we find that the planets do not every night exactly retain their relative places among the stars, we next admit that such deviations from the law, which all the rest seem to obey, are owing to a proper motion of their own. To apply this to the solar system. – Astronomers have already observed what they call a proper motion in several of the fixed stars, and the same may be supposed of them all. We ought, therefore, to resolve that which is common to all the stars, which are found to have what has been called a proper motion, into a single real motion of the solar system, as far as that will answer the known facts, and only to attribute to the proper motion of each particular star the deviations from the general law the stars seem to follow in those movements.

By Dr. MASKELYNE's account of the proper motion of some principal stars,[6] we find that Sirius, Castor, Procyon, Pollux, Regulus, Arcturus, and α Aquilae, appear to have respectively the following proper motions in right ascension: – 0″,63; –0″,28 ; –0″,80; –0″,93; –0″,41; –1″,46; and +0″,57; and two of them, Sirius and Arcturus, in declination, *viz.* 1″,20 and 2″,01, both southward. Let fig. 3. represent an equatorial zone, with the above mentioned stars referred to it, according to their respective right ascensions, having the solar system in its center. Assume the direction AB from a point somewhere not far from the 77th degree of right ascension to its opposite 257th degree, and suppose the sun to move in that direction from S towards B; then will that one motion answer that of all the stars together: for if the supposition be true, Arcturus, Regulus, Pollux, Procyon, Castor, and Sirius, should appear to decrease in right ascension, while α Aquila, on the contrary, should appear to increase. Moreover, suppose the sun to ascend at the same time in the same direction towards some point in the northern hemisphere, for instance, towards the constellation of Hercules; then will also the observed change of declination of Sirius and Arcturus be resolved into the single motion of the solar system. I am well aware of the many yet remaining difficulties, such as the correspondence of the exact quantity of each star's observed proper motion with the quantity that will be assigned to it by this hypothesis; but we ought to remember, that the very different and still unknown relative distances of the fixed stars must, for a good while yet, leave us in the dark about the particular and strict application of the theory; and that any deviation

[6] Astronomical Observations made at the Royal Observatory at Greenwich [**1** (London, 1776)].

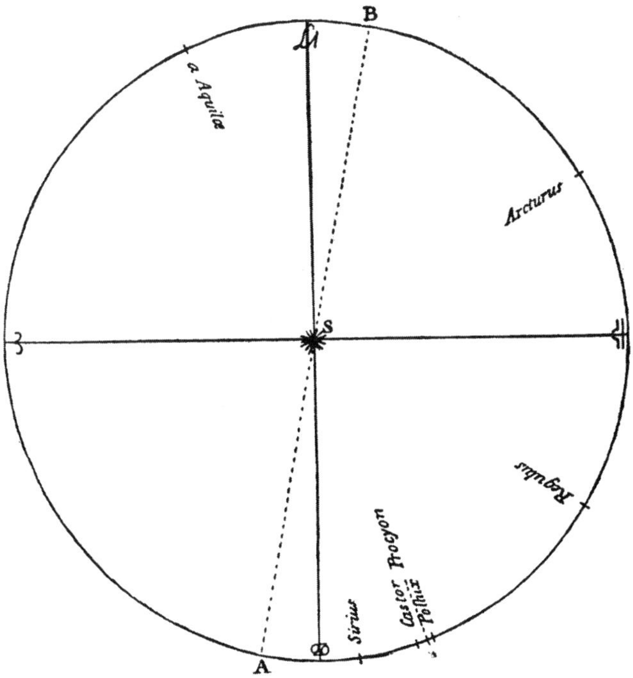

Fig. 3

from it may easily be accounted for by the still unknown *real* proper motion of the stars: for if the solar system have the motion I ascribe to it, then what astronomers have already observed concerning the change of place of the stars, and have called their proper motion, will become only an *apparent* motion; and it will still be left to future observations to point out, by the deviations from the general law which the stars will follow in those apparent motions, what may be their real proper motions as well as relative distances. But lest I should be censured for admitting so new and capital a motion upon too slight a foundation, I must observe, that the concurrence of those seven principal stars cannot but give some value to an hypothesis that will simplify the celestial motions in general. We know that the sun, at the distance of a fixed star, would appear like one of them; and from analogy we conclude the stars to be suns. Now, since the apparent motions of these seven stars may be accounted for, either by supposing them to move just in the manner they appear to do, or else by supposing the sun alone to have a motion in a direction, somehow not far from that which I have assigned to it, I think we are no more authorised to suppose the sun at rest than we should be to deny the diurnal motion of the earth, except in this respect, that the proofs of the latter are very numerous, whereas the former rests only on a few though capital testimonies. But to proceed: I have only mentioned the motions of those seven principal stars, as being the most noticed and best ascertained of all; I will now adduce a farther confirmation of the same from other stars.

M. DE LA LANDE gives us the following table of the proper motion of 12 stars, both in right ascension and declination, in 50 years.[7]

Table 1

Etoiles.	Chang. d'asc. droite.		Chang. de déclinaison.	
	′	″	′	″
Arcturus	−1	11	−1	55
Sirius	−	37	−	52
β Cygni	−	3	+	49
Procyon	−	33	−	47
ε Cygni	+	20	+	34
γ Arietis	−	14	−	29
γ Gemin.	−	8	−	24
Aldébaran	+	3	−	18
β Gemin.	−	48	−	16
γ Piscium	+	53	+	7
α Aquilæ	+	32	−	4
α Gemin.	−	24	−	1

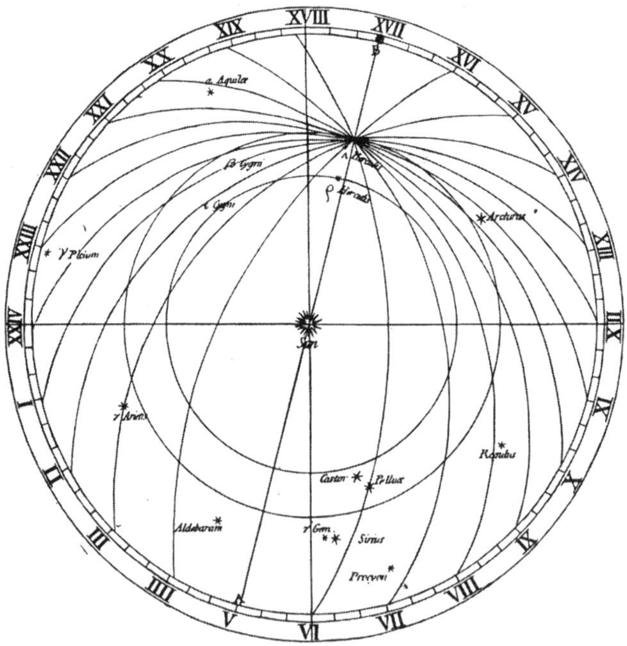

Fig. 4

[7] Ast. par M. DE LA LANDE, tom. IV. p. 685.

Fig. 4. represents them projected on the plane of the equator. They are all in the northern hemisphere, except Sirius, which must be supposed to be viewed in the concave part of the opposite half of the globe, while the rest are drawn on the convex surface. Regulus being added to that number, and Castor being double, we have 14 stars. Every star's motion, except Regulus, is assigned in declination as well as in right ascension, so that we have no less than 27 motions given to account for. Now, by assuming a point somewhere near λ Herculis, and supposing the sun to have a proper motion towards that part of the heaven, we shall satisfy 22 of these motions.[8] For β Cygni, α Aquilae, ε Cygni, γ Piscium, γ Arietis, and Aldebaran, ought, upon the supposed motion of the sun, to have an apparent progression, according to the hour circle XVIII, XIX, XX, &c. or to increase in right ascension, while Arcturus, Regulus, the two stars of α Geminorum, Pollux, Procyon, Sirius, and γ Geminorum, should apparently go back in the order XVI, XV, XIV, &c. of the hour circle, so as to decrease in right ascension; but according to M. DE LA LANDE's table, excepting β Cygni and γ Arietis, all these motions really take place. With regard to the change of declination, we see that every star in the table should go towards the south;[9] and here we find but three exceptions in β and ε Cygni, and γ Piscium; so that upon the whole we have but five deviations out of 27 known motions which this hypothesis will not account for. And these exceptions must be resolved into the real proper motion of the stars.

There are also some very striking circumstances in the quantities of these motions that deserve our notice. First, Arcturus and Sirius being the largest of the stars, and therefore probably the nearest, ought to have the most apparent motion, both in right ascension and declination, which is agreeable to observation, as we find by the table. Next, in regard to the right ascension only, Arcturus being better situated to shew its motion, by theorem 2 [p. 90], ought to have it much larger, which we find it has. Aldebaran, both badly situated and considerably smaller than the two former, by the same theorem ought to shew but little motion. Procyon, better situated than Sirius, though not quite so large, should have almost as much motion; for by the third theorem, on supposing it farther off because it appears smaller, the effect of the sun's motion will be lessened upon it; whereas, on the other hand, by the

[8] [In his original version, Herschel thought that he could account also for the change in declination of β and ε Cygni, so satisfying no fewer than twenty-four out of the twenty-seven motions and leaving only three "deviations".]

[9] [This sentence originally read: "With regard to the change of declination, we see that β and ε Cygni should go towards the north pole as lying between the Pole and the point towards which the sun is supposed to move; and that all the rest ought to go towards the south, or encrease their n[orth] polar distance."]

second theorem, its better situation will partly compensate for its greater distance. This again is conformable to the table. ε Cygni very favourably situated, though but a small star, should shew it considerably as well as α Aquilae; whereas β Cygni should have but little motion: and γ Piscium, best situated of all, should have a great increase of right ascension, and these deductions also agree with the table.

In the last place, a very striking agreement with the hypothesis is displayed in Castor and Pollux. They are both pretty well situated, and we accordingly find that Pollux, for the size of the star, shews as much motion in right ascension as we could expect; but it is remarkable, and seemingly contrary to our hypothesis, that Castor, equally well placed, shews by the table no more than one half of the motion of Pollux. Now, if we recollect that the former is a double star, consisting of two stars not much different in size, we can allow but about half the light to each of them, which affords a strong presumption of their being at a greater distance, and therefore their partial systematical parallax, by the third theorem, ought to be so much less than that of Pollux, which agrees wonderfully with observation.[10] Not to mention the great difficulty in which we should be involved, were we to suppose the motion of Castor to be really in the star: for how extraordinary must appear the concurrence, that two stars, namely those that make up this apparently single star, should both have a proper motion so exactly alike, that in all our observations hitherto they have not been found to disagree a single second, either in right ascension or declination, for fifty years together! Does not this seem strongly to point out the common cause, the motion of the solar system?

With respect to the change of declination I would observe, that the point of λ Herculis, which in fig. 4. is assumed as the Apex[11] of the solar motion is not perhaps the best selected. A somewhat more northern situation may agree better with the changes of declination of Arcturus and Sirius, which capital

[10] If the light of Castor was exactly equal to that of Pollux, and the two stars, which make up the former star, were perfectly of the same size, we might, on that account, suppose the distance of Castor from us to be to that of Pollux as $\sqrt{2} : 1$; but Castor is in fact something less bright; and this consideration, added to the former, will make it probable enough that its distance may perhaps be double that of Pollux. [This note was added after the paper was read. Maskelyne suggested in a letter of 15 March 1783 (RAS W.1/13. M.22), that the stars of Castor "may belong to a particular system, which of course would all move together". Herschel in his reply of 17 March agreed, but argued that until this was proved to be so the evidence added to the probabilities in favour of his case.]

[11] I use the term Apex here to denote that point of fig. 4., wherein all great circles, drawn through the supposed direction of the motion of the solar system, intersect, and which, in other stereographic projections, is generally a pole, either of the ecliptic or equator. As this point is the northern or elevated hemisphere, the sun, by tending to it, may be said to ascend, and the term Apex may perhaps not be an improper one.

stars may perhaps be the most proper to lead us in this hypothesis; but as we should be guided by facts in researches of this nature, it may be as well to expect the assistance of future observations before we are too particular in determining this point.[12]

It may be expected I should also mention something concerning the quantity of the solar motion; but here I can only offer a few distant hints. From the annual parallax of the fixed stars, which, from my own observations, I find much less than it has hitherto been proved to be, we may certainly admit (without entering into a subject which I reserve for a future opportunity) that the diameter of the earth's orbit, at the distance of Sirius or Arcturus, would not nearly subtend an angle of one second; but the apparent motion of Arcturus, if owing to a translation of the solar system, amounts to no less than 2″,7 a year, as will appear if we compound the two motions of 1′ 11″ in right ascension, and 1′ 55″ in declination, into one single motion, and reduce it to an annual quantity. Hence we may in a general way estimate, that the solar motion can certainly not be less than that which the earth has in her annual orbit.

I have now only to add, that it is to be expected future observations will soon throw more light upon this interesting subject, and either fully establish or overturn the hypothesis of the motion of the whole solar system. To this end I have already begun a series of observations upon several zones of double stars; and should the result of them be against these conjectures, I shall be the first to endeavour to point out the fallacy of them.

[WILLIAM HERSCHEL.]

Datchet near Windsor,
Feb. 1, 1783.

[*Instead of as above, p. 88, line 42 to p. 89 line 17, Herschel's paper as originally submitted read:* "... less in motion? May not a star which is lost be looked upon as one whose proper motion has carried it directly out of our sight? A new one which appears may be such a one whose proper motion brings it immediately into our view: A star which increases in magnitude and changes its place by slow degrees, may be one which moves in an oblique direction towards us: A gradually vanishing star, with change of place, may be that which moves obliquely from us: And a star that changes its place without altering its magnitude may be such a one whose motion is at rectangles to a line drawn from the Sun to that

[12] From the additional testimony of other capital stars considered in the postscript [not reproduced] it now appears, that the point of λ Herculis is probably as well chosen as any we can fix upon in that part of the heavens. [Footnote added after the paper was read. The original version of Herschel's text of this passage is reproduced on p. 98.]

star. Now if observations continually point out such changes in the apparent magnitudes and reciprocal distances of the stars who can hesitate to pronounce their proper motions? And if the proper motion. . .”

Instead of as above, p. 96, line 24 to p. 97 line 4, Herschel's paper as originally submitted read: “If the change in declination of β & ε Cygni be thrown on their own proper motion, rather than that of the Solar-System, we shall be able to make other quantities agree much better with observation, by supposing the Sun to move towards a more nothern star, such as ρ Herculis. It seems the motion of Arcturus and Sirius is too great to permit the Apex to lie so low as either λ or the 51st Herculis. If it were raised to the parallel of the 50th of the same constellation, we should only throw out β Cygni, which is a star from other reasons much to be suspected of a real proper motion; but even this elevation would hardly suffice: I would rather consider Arcturus and Sirius as the leading marks to regulate that point. It can not be said that I act arbitrarily in assigning to ε and β Cygni motions at pleasure, as best suits my hypothesis; for since proper motions have actually been observed, if they are not to be accounted for from that of the Sun, they must needs be in the stars themselves; and as in this hypothesis I would intirely be directed by facts, it is highly proper, where contrary motions appear, to lean to the side which has the strongest arguments in its favour.”]

1784: "Observations tending to investigate the construction of the heavens"

Summary: "I have also sometime since given a paper called 'An account of some Observations tending to investigate the Construction of the Heavens.' In it I have shewn that the heavens contain Strata of stars and of nebulae, and that the Sun is situated in a stratum of stars. By the place which I assign to the Sun in this stratum as the most probable I am enabled to account for all the phenomena of the Via Lactea, which is evidently composed of stars so close as to contain about 50 thousand of them within the space of 15 degrees long by two degrees broad, large enough to be distinctly counted with my new great 20 feet reflector. The same situation of the Sun will point out to us the probable cause of the motion of the solar system.

"The same instrument which has been mentioned has resolved most of the nebulae contained in the Connoissance des temps for 1783 & 1784, into stars and shews that they are clusters of them. When I point out the strata of nebulae, (one of which is so rich that in passing thro' a section of it I detected not less than 31 new nebulae in only 36 minutes of time) I shew that they contain the greatest variety of them, and having already seen 466 new ones, never before observed, I give a delineation of many of them. I have also joined a method for pursuing the investigation of the sidereal & nebulous strata."

Herschel to G. C. Lichtenberg, 20 May 1784.[1]

Account of some Observations tending to investigate the Construction of the Heavens[2]

Read June 17, 1784.

IN a former paper I mentioned, that a more powerful instrument was preparing for continuing my reviews of the heavens. The telescope I have lately completed,

[1] [RAS W.1/1, 111–112.] [2] [*Phil. Trans.*, **74** (1784), 437–451.]

though far inferior in size to the one I had undertaken to construct when that paper was written, is of the Newtonian form, the object speculum being of 20 feet focal length, and its aperture 18 7/10 inches. The apparatus on which it is mounted is contrived so as at present to confine the instrument to a meridional situation, and by its motions to give the right-ascension and declination of a celestial object in a coarse way; which, however, is sufficiently accurate to point out the place of the object, so that it may be found again. It will not be necessary to enter into a more particular description of the apparatus, since the account I have now the honour of communicating to the Royal Society regards rather the performance of the telescope than its construction.

It would, perhaps, have been more eligible to have waited longer, in order to complete the discoveries that seem to lie within the reach of this instrument, and are already, in some respects, pointed out to me by it. By taking more time I should undoubtedly be enabled to speak more confidently of the *interior construction* of the heavens, and its various *nebulous and sidereal strata* (to borrow a term from the natural historian) of which this paper can as yet only give a few outlines, or rather hints. As an apology, however, for this prematurity, it may be said, that the end of all discoveries being communication, we can never be too ready in giving facts and observations, whatever we may be in reasoning upon them.

Hitherto the sidereal heavens have, not inadequately for the purpose designed, been represented by the concave surface of a sphere, in the center of which the eye of an observer might be supposed to be placed. It is true, the various magnitudes of the fixed stars even then plainly suggested to us, and would have better suited the idea of an expanded firmament of three dimensions; but the observations upon which I am now going to enter still farther illustrate and enforce the necessity of considering the heavens in this point of view. In future, therefore, we shall look upon those regions into which we may now penetrate by means of such large telescopes, as a naturalist regards a rich extent of ground or chain of mountains, containing strata variously inclined and directed, as well as consisting of very different materials. A surface of a globe or map, therefore, will but ill delineate the interior parts of the heavens.

It may well be expected, that the great advantage of a large aperture would be most sensibly perceived with all those objects that require much light, such as the very small and immensely distant fixed stars, the very faint nebulae, the close and compressed clusters of stars, and the remote planets.

On applying the telescope to a part of the *via lactea*, I found that it completely resolved the whole whitish appearance into small stars, which my former telescopes had not light enough to effect. The portion of this extensive tract, which it has hitherto been convenient for me to observe, is that immediately about the hand and club of Orion. The glorious multitude of stars of all possible sizes that presented themselves here to my view was truly astonishing; but, as the dazzling brightness of glittering stars may easily mislead us so far as to estimate their number greater than it really is, I endeavoured to ascertain this point by counting many fields, and computing, from a mean of them, what a certain given portion of the milky way might contain. Among many trials of this sort I found, last January the 18th, that six fields, promiscuously taken, contained 110, 60, 70, 90, 70, and 74 stars each. I then tried to pick out the most vacant place that was to be found in that neighbourhood, and counted 63 stars. A mean of the first six gives 79 stars for each field. Hence, by allowing 15 [arc]minutes of a great circle for the diameter of my field of view, we gather, that a belt of 15 degrees long and two broad, or the quantity which I have often seen pass through the field of my telescope in one hour's time, could not well contain less than fifty thousand stars, that were large enough to be distinctly numbered. But, besides these, I suspected at least twice as many more, which, for want of light, I could only see now and then by faint glittering and interrupted glimpses.

The excellent collection of nebulae and clusters of stars which has lately been given in the *Connoissance des Temps* for 1783 and 1784, leads me next to a subject which, indeed, must open a new view of the heavens. As soon as the first of these volumes came to my hands, I applied my former 20-feet reflector of 12 inches aperture to them; and saw, with the greatest pleasure, that most of the nebulae, which I had an opportunity of examining in proper situations, yielded to the force of my light and power, and were resolved into stars. For instance, the 2d, 5, 9, 10, 12, 13, 14, 15, 16, 19, 22, 24, 28, 30, 31, 37, 51, 52, 53, 55, 56, 62, 65, 66, 67, 71, 72, 74, 92, all which are said to be nebulae without stars, have either plainly appeared to be nothing but stars, or at least to contain stars, and to shew every other indication of consisting of them entirely.[3] I have examined them with a careful scrutiny of various powers and light, and generally in the meridian. I should mention, that five of the above, *viz.* the 16th, 24, 37, 52, 67, are called clusters of stars containing nebulosity; but my instrument resolving also that portion of them which is called nebulous into stars of a much smaller size, I have placed them into the above number. To these may be added the 1st, 3d, 27, 33, 57, 79, 81, 82, 101, which in my 7, 10, and 20-feet reflectors shewed a mottled

[3] [Of these twenty-nine Messier objects, eighteen are globular clusters and six galactic clusters. The successful resolution into stars of many of the globular clusters in Messier's catalogue is a clear indication of the quality of Herschel's telescopes. But Herschel was overoptimistic in expecting M31, 51, 65, 66, and 74 to be visually resolved into stars with a further increase in power: they are galaxies and much too distant for Herschel's instrument to detect individual stars. – D.]

kind of nebulosity, which I shall call resolvable;[4] so that I expect my present telescope will, perhaps, render the stars visible of which I suppose them to be composed. Here I might point out many precautions necessary to be taken with the very best instruments, in order to succeed in the resolution of the most difficult of them; but reserving this at present too extensive subject for a future opportunity, I proceed to speak of the effects of my last instrument with regard to nebulae.

My present pursuits, as I observed before, requiring this telescope to act as a fixed instrument, I found it not convenient to apply it to any other of the nebulae in the *Connoissance des Temps* but such as came in turn; nor, indeed, was it necessary to take any particular pains to look for them, it being utterly impossible that any one of them should escape my observation when it passed the field of view of my telescope. The few which I have already had an opportunity of examining, shew plainly that those most excellent French astronomers, Mess. MESSIER and MECHAIN, saw only the more luminous part of their nebulae; the feeble shape of the remainder, for want of light, escaping their notice. The difference will appear when we compare my observation of the 98th nebula[5] with that in the *Connoissance des Temps* for 1784, which runs thus: "Nébuleuse sans étoile, d'une lumière extrêmement foible, au dessus de l'aile boréale de la Vierge, sur le parallèle et près de l'étoile N° 6, cinquième grandeur, de la chevelure de Bérénice, suivant FLAMSTEED. M. MECHAIN la vit le 15 Mars, 1781." My observation of the 30th of December, 1783, is thus: A large, extended, fine nebula. Its situation shews it to be M. MESSIER's 98th; but from the description it appears, that that gentleman has not seen the whole of it, for its feeble branches extend above a quarter of a degree, of which no notice is taken. Near the middle of it are a few stars visible, and more suspected. My field of view will not quite take in the whole nebula.[6] See fig. 1. Again, N° 53,[7] "Nébuleuse sans étoiles, découverte au-dessous et près de la chevelure de Bérénice, à peu de distance de l'étoile quarante-deuxième de cette constellation, suivant FLAMSTEED. Cette nébuleuse est ronde et apparente, &c." My observation of the 170th Sweep runs thus: A cluster of very close stars; one of the most beautiful objects I remember to have seen in the heavens. The cluster appears under the form of a solid ball, consisting of small stars, quite compressed into one blaze of light, with a great number of loose ones surrounding it, and distinctly visible in the general mass. See fig. 2.

4 [Of the nine Messier objects, only M3 and M79 are globular clusters. The "mottled kind of nebulosity, which I shall call resolvable" misled Herschel in M33, 81, 82, and 101 (galaxies), whilst M1, 27, and 57 are gaseous nebulae (M1, the "Crab" nebula, the remnant of the supernova of A.D. 1054; M27 and 57, planetary nebulae). – D.]

5 [M98 = NGC 4192, a giant edge-on spiral in the Virgo cluster of galaxies. – D.]

6 [This is surprising as the extent of M98 is only 10′ and Herschel's normal field of view was over 15′. – S.]

7 [M53 = NGC 5024, a globular cluster well described by Herschel. – D.]

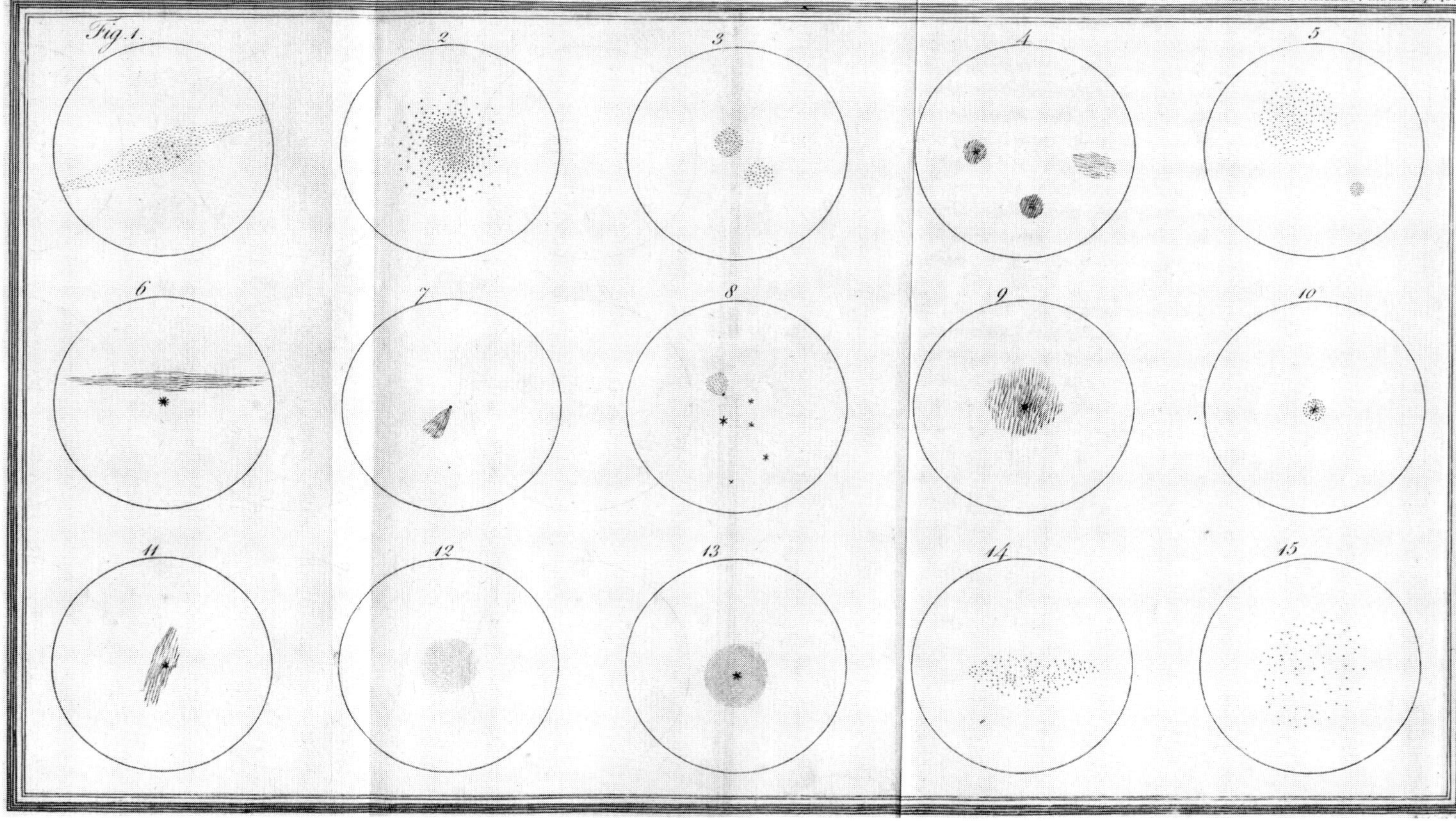
Fig. 1.
2
3
4
5
6
7
8
9
10
11
12
13
14
15

When I began my present series of observations, I surmised, that several nebulae might yet remain undiscovered, for want of sufficient light to detect them; and was, therefore, in hopes of making a valuable addition to the clusters of stars and nebulae already collected and given us in the work before referred to, which amount to 103. The event has plainly proved that my expectations were well founded: for I have already found 466 new nebulae and clusters of stars, none of which, to my present knowledge, have been seen before by any person; most of them, indeed, are not within the reach of the best common telescopes now in use. In all probability many more are still in reserve; and as I am pursuing this track, I shall make them up into separate catalogues, of about two or three hundred at a time, and have the honour of presenting them in that form to the Royal Society.

A very remarkable circumstance attending the nebulae and clusters of stars is, that they are arranged into strata, which seem to run on to a great length; and some of them I have already been able to pursue, so as to guess pretty well at their form and direction. It is probable enough, that they may surround the whole apparent sphere of the heavens, not unlike the milky way, which undoubtedly is nothing but a stratum of fixed stars. And as this latter immense starry bed is not of equal breadth or lustre in every part, nor runs on in one straight direction, but is curved and even divided into two streams along a very considerable portion of it; we may likewise expect the greatest variety in the strata of the clusters of stars and nebulae. One of these nebulous beds is so rich, that, in passing through a section of it, in the time of only 36 minutes, I detected no less than 31 nebulae, all distinctly visible upon a fine blue sky. Their situation and shape, as well as condition, seem to denote the greatest variety imaginable. In another stratum, or perhaps a different branch of the former, I have seen double and treble nebulae, variously arranged; large ones with small, seeming attendants; narrow but much extended, lucid nebula or bright dashes; some of the shape of a fan, resembling an electric brush, issuing from a lucid point; others of the cometic shape, with a seeming nucleus in the center; or like cloudy stars, surrounded with a nebulous atmosphere; a different sort again contain a nebulosity of the milky kind, like that wonderful, inexplicable phaenomenon about θ Orionis; while others shine with a fainter, mottled kind of light, which denotes their being resolvable into stars.[8] See fig. 3, &c. But it would be too extensive at present to enter more minutely into such circumstances, therefore I proceed with the subject of nebulous and sidereal strata.

It is very probable, that the great stratum, called the milky way, is that in which the sun is placed, though perhaps not in the very center of its thickness.

[8] [Herschel is here describing a wide variety of objects: both areas of sky in high galactic latitude containing many galaxies of different forms, and also ("that ... inexplicable phaenomenon about θ Orionis", i.e. the Orion Nebula) galactic nebulae. – D.]

We gather this from the appearance of the Galaxy, which seems to encompass the whole heavens, as it certainly must do if the sun is within the same. For, suppose a number of stars arranged between two parallel planes, indefinitely extended every way, but at a given considerable distance from each other; and, calling this a sidereal stratum, an eye placed somewhere within it will see all the stars in the direction of the planes of the stratum projected into a great circle, which will appear lucid on account of the accumulation of the stars; while the rest of the heavens, at the sides, will only seem to be scattered over with constellations, more or less crowded, according to the distance of the planes or

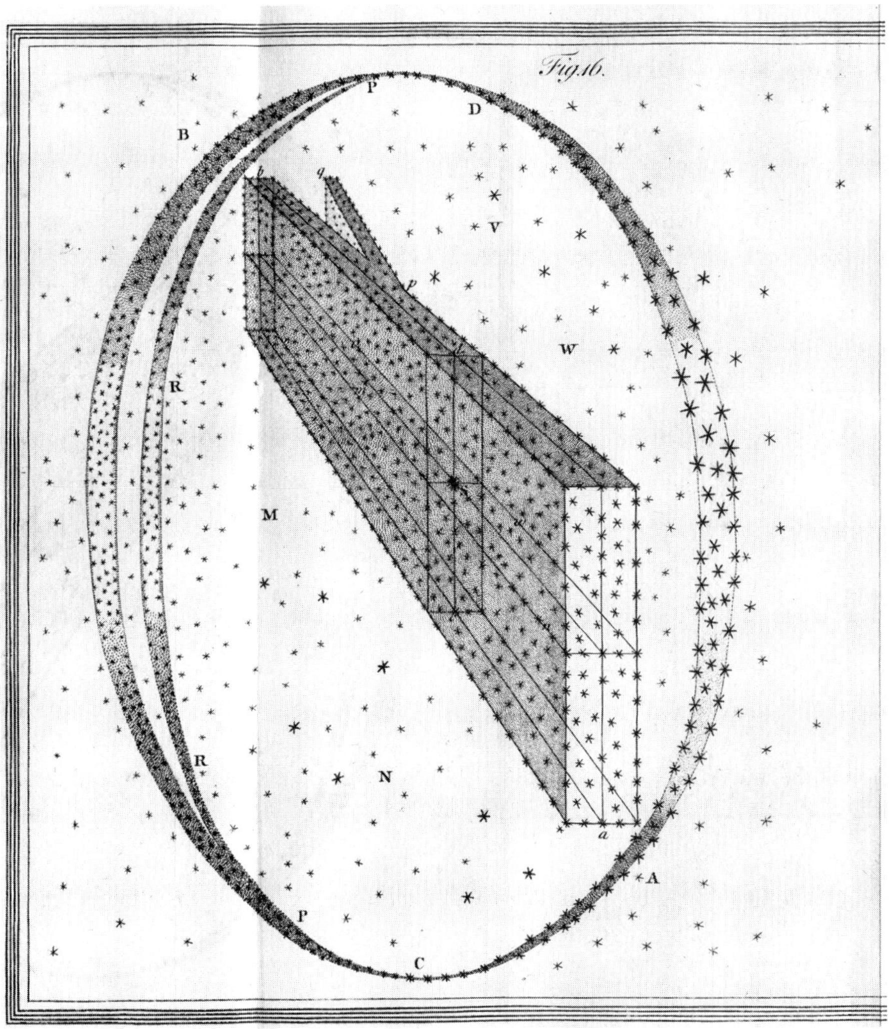

number of stars contained in the thickness or sides of the stratum. Thus, in fig. 16 an eye at S within the stratum *ab*, will see the stars in the direction of its length *ab*, or height *cd*, with all those in the intermediate situations, projected into the lucid circle ACBD; while those in the sides *my*, *nw*, will be seen scattered over the remaining part of the heavens at MVNW.

If the eye were placed somewhere without the stratum, at no very great distance, the appearance of the stars within it would assume the form of one of the less circles of the sphere, which would be more or less contracted to the distance of the eye; and if this distance were exceedingly increased, the whole stratum might at last be drawn together into a lucid spot of any shape, according to the position, length, and height of the stratum.

Let us now suppose, that a branch, or smaller stratum, should run out from the former, in a certain direction, and let it also be contained between two parallel planes extended indefinitely onwards, but so that the eye may be placed in the great stratum somewhere before the separation, and not far from the place where the strata are still united. Then will this second stratum not be projected into a bright circle like the former, but will be seen as a lucid branch proceeding from the first, and returning to it again at a certain distance less than a semi-circle.

Thus, in the same figure, the stars in the small stratum *pq* will be projected into a bright arch at PRRP, which, after its separation from the circle CBD, unites with it again at P.

What has been instanced in parallel planes may easily be applied to strata irregularly bounded, and running in various directions; for their projections will of consequence vary according to the quantities of the variations in the strata and the distance of the eye from the same. And thus any kind of curvatures, as well as various different degrees of brightness, may be produced in the projections.

From appearances then, as I observed before, we may infer, that the sun is most likely placed in one of the great strata of the fixed stars, and very probably not far from the place where some smaller stratum branches out from it. Such a supposition will satisfactorily, and with great simplicity, account for all the phaenomena of the milky way, which, according to this hypothesis, is no other than the appearance of the projection of the stars contained in this stratum and its secondary branch. As a farther inducement to look on the Galaxy in this point of view, let it be considered, that we can no longer doubt of its whitish appearance arising from the mixed lustre of the numberless stars that compose it. Now, should we imagine it to be an irregular ring of stars, in the center nearly of which we must then suppose the sun to be placed, it will appear not a little extraordinary, that the sun, being a fixed star like those which compose this imagined ring, should just be in the center of such a multitude of celestial bodies, without any apparent reason for this singular distinction; whereas, on our supposition, every star in this stratum, not very near the termination of its length or height, will be so placed as also to have its own Galaxy, with only such variations in the form and lustre of it, as may arise from the particular situation of each star.

Various methods may be pursued to come to a full knowledge of the sun's place in the sidereal stratum, of which I shall only mention one as the most general and most proper for determining this important point, and which I have already begun to put in practice. I call it *Gaging the Heavens*, or the *Star-Gage*.[9] It consists in repeatedly taking the number of stars in ten fields of view of my reflector very near each other, and by adding their sums, and cutting off one decimal on the right, a mean of the contents of the heavens, in all the parts which are thus gaged, is obtained. By way of example, I have joined a short table, extracted from the gages contained in my journal, by which it appears, that the number of stars increases very fast as we approach the Via Lactea.

Thus, in the parallel from 92 to 94 degrees north polar distance, and R.A. 15 h. 10′, the star-gage runs up from 9.4 stars in the field to 18.6 in about an hour and a half; whereas in the parallel from 78° to 80° north polar distance, and R.A. 11, 12, 13, and 14 hours, it very seldom rises above 4. We are, however, to remember, that with different instruments the account of the gages will be very different, especially on our supposition of the situation of the sun in a stratum of stars. For, let *ab*, fig. 17, be the stratum, and suppose the small circle *ghlk* to represent the space into which, by the light and power of a given telescope, we

N.P.D. 92 to 94°.			N.P.D. 78 to 80°.		
R.A.		Gage.	R.A.		Gage.
15	10	9.4	11	16	3.1
15	22	10.6	12	31	3.4
15	47	10.6	12	44	4.6
16	8	12.1	12	49	3.9
16	25	13.6	13	5	3.8
16	37	18.6	14	30	3.6

[9] [A pioneering exercise in the collection and interpretation of stellar statistics.]

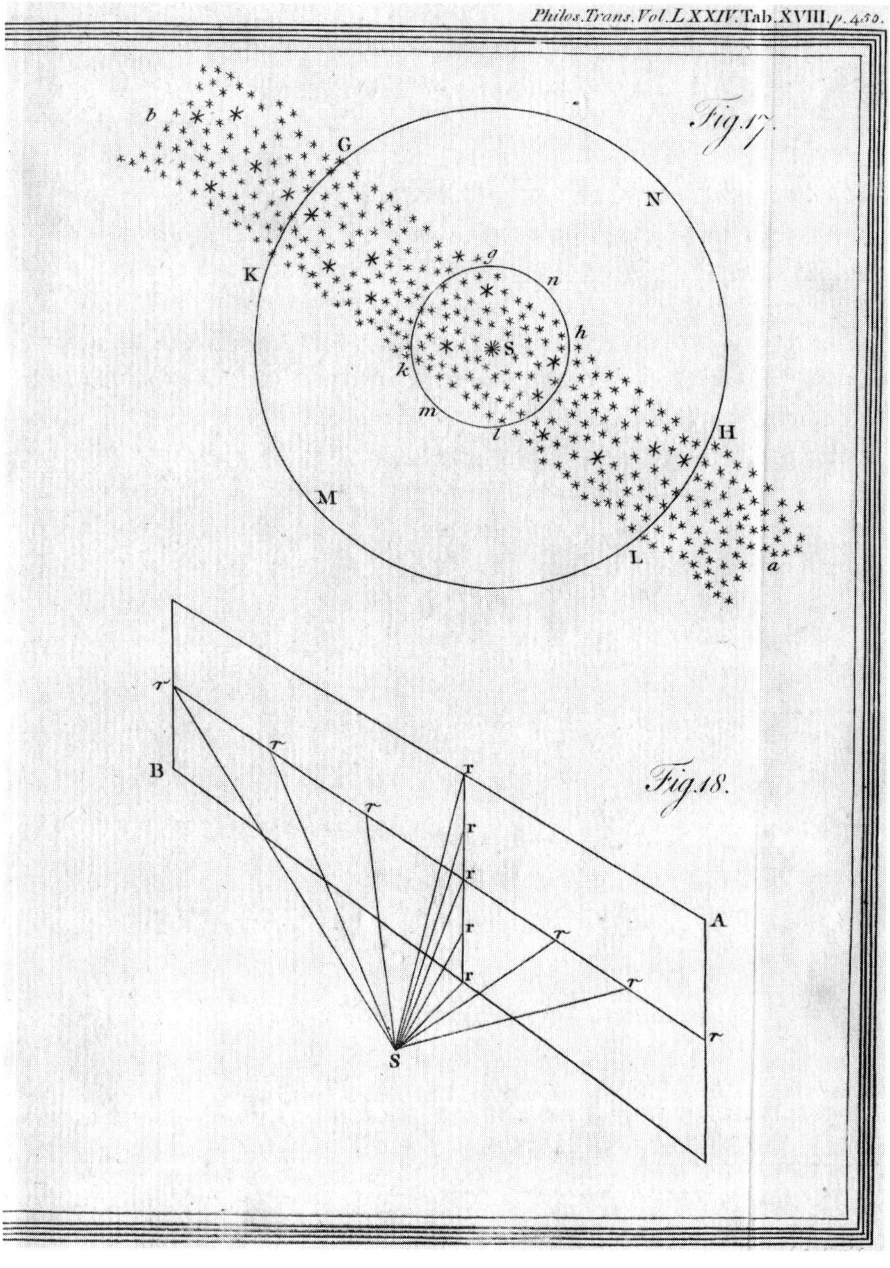

may penetrate; and let GHLK be the extent of another portion, which we are enabled to visit by means of a larger aperture and power; it is evident, that the gages with the latter instrument will differ very much in their account of stars contained at MN, and at KG or LH; when with the former they will hardly be affected by the change from *mn* to *kg* or *lh*. And this accounts for what a celebrated author says concerning the effects of telescopes, by which we must understand the best of those that are in common use.[10]

It would not be safe to enter into an application of these, and such other gages as I have already taken, till they are sufficiently continued and carried all over the heavens. I shall, therefore, content myself with just mentioning that the situation of the sun will be obtained, from considering in what manner the star-gage agrees with the length of a ray revolving in several directions about an assumed point, and cut off by the bounds of the stratum. Thus, in fig. 18, let S be the place of an observer; Srrr, *Srrr*, lines in the planes rSr, *rSr*, drawn from S within the stratum to one of the boundaries, here represented by the plane AB. Then, since neither the situation of S, nor the form of the limiting surface AB, is given, we are to assume a point, and apply to it lines proportional to the several gages that have been obtained, and at such angles from each other as they may point out; then will the termination of these lines delineate the boundary of the stratum, and consequently manifest the situation of the sun within the same. But to proceed.

If the sun should be placed in the great sidereal stratum of the milky way, and, as we have surmised above, not far from the branching out of a secondary stratum, it will very naturally lead us to guess at the cause of the probable motion of the solar system: for the very bright, great node of the Via Lactis, or union of the two strata about Cepheus and Cassiopeia, and the Scorpion and Sagittarius, points out a conflux of stars manifestly quite sufficient to occasion a tendency towards that node in any star situated at no very great distance; and the secondary branch of the Galaxy not being much less than a semi-circle seems to indicate such a situation of our solar system in the great undivided stratum as the most probable.

What has been said in a former paper [above, pp. 87–97] on the subject of the solar motion seems also to support this supposed situation of the sun; for the apex there assigned lies nearly in the direction of a motion of the sun towards the node of the strata. Besides, the joining stratum making a pretty large angle

[10] On voit avec les télescopes des étoiles dans toutes les parties du ciel, à peu près comme dans la voie lactée, ou dans les nébuleuses. On ne sauroit douter qu'une partie de l'éclat et de la blancheur de la voie lactée, ne provienne de la lumière des petites étoiles qui s'y trouvent en effet par millions; cependant, avec les plus grands télescopes, on n'en distingue pas assés, et elles n'y sont pas assés rapprochées les unes des autres pour qu'on puisse attribuer à celles qu'on distingue la blancheur de la voie lactée, si sensible à la vue simple. L'on ne sauroit donc prononcer que les étoiles soient la seule cause de cette blancheur, quoique nous ne connoissions aucune manière satisfaisante de l'expliquer. *Ast.* M. DE LA LANDE, § 833.

at the junction with the primary one, it may easily be admitted, that the motion of a star in the great stratum, especially if situated considerably towards the side farthest from the small stratum, will be turned sufficiently out of the straight direction of the great stratum towards the secondary one. But I find myself insensibly led to say more on this subject than I am as yet authorised to do; I will, therefore, return to those observations which have suggested the idea of celestial strata.

In my late observations on nebulae I soon found, that I generally detected them in certain directions rather than in others; that the spaces preceding them were generally quite deprived of their stars, so as often to afford many fields without a single star in it; that the nebulae generally appeared some time after among stars of a certain considerable size, and but seldom among very small stars; that when I came to one nebula, I generally found several more in the neighbourhood; that afterwards a considerable time passed before I came to another parcel; and these events being often repeated in different altitudes of my instrument, and some of them at a considerable distance from each other, it occurred to me, that the intermediate spaces between the sweeps might also contain nebulae; and finding this to hold good more than once, I ventured to give notice to my assistant at the clock, "to prepare, since I expected in a few minutes to come at a stratum of the nebulae, finding myself already" (as I then figuratively expressed it) "on nebulous ground." In this I succeeded immediately; so that I now can venture to point out several not far distant places, where I shall soon carry my telescope, in expectation of meeting with many nebulae. But how far these circumstances of vacant places preceding and following the nebulous strata, and their being as it were contained in a bed of stars, sparingly scattered between them, may hold good in more distant portions of the heavens, and which I have not yet been able to visit in any regular manner, I ought by no means to hazard a conjecture. The subject is new, and we must attend to observations, and be guided by them, before we form general opinions.

Before I conclude, I may, however, venture to add a few particulars about the direction of some of the capital strata or their branches. The well known nebula of Cancer,[11] visible to the naked eye, is probably one belonging to a certain stratum, in which I suppose it to be so placed as to lie nearest to us. This stratum

[11] ["well known nebula of Cancer" = M44, Praesepe, a nearby galactic star cluster resolvable with the slightest optical aid, as Herschel well knew. What Herschel had in mind as the "stratum of Cancer" is not very clear: probably the association of M44 and M67 (another galactic cluster) with the numerous faint naked-eye stars that make up the constellation of Cancer, stretching down to Hydra. With M44 they give this part of the sky a hazy appearance at a first glance. There is no unusually great number of galaxies here. – D.]

I shall call that of Cancer.[12] It runs from ε Cancri towards the south over the 67 nebula of the *Connoissance des Temps*, which is a very beautiful and pretty much compressed cluster of stars, easily to be seen by any good telescope, and in which I have observed above 200 stars at once in the field of view of my great reflector, with a power of 157. This cluster appearing so plainly with any good, common telescope, and being so near to the one which may be seen by the naked eye, denotes it to be probably the next in distance to that[13] within the quartile formed by γ, δ, η, θ; from the 67th nebula the stratum of Cancer proceeds towards the head of Hydra; but I have not yet had time to trace it farther than the equator.

Another stratum, which perhaps approaches nearer to the solar system than any of the rest, and whose situation is nearly at rectangles to the great sidereal stratum in which the sun is placed, is that of Coma Berenices,[14] as I shall call it.[15] I suppose the Coma itself to be one of the clusters in it, and that, on account of its nearness, it appears to be so scattered. It has many capital nebulae very near it; and in all probability this stratum runs on a very considerable way. It may, perhaps, even make the circuit of the heavens, though very likely not in one of the great circles of the sphere: for, unless it should chance to intersect the great sidereal stratum of the milky way before-mentioned, in the very place in which the sun is stationed, such an appearance could hardly be produced. However, if the stratum of Coma Berenices should extend so far as (by taking in the assistance of M. MESSIER's and M. MECHAIN's excellent observations of scattered nebulae, and some detached former observations of my own) I apprehend it may, the direction of it towards the north lies probably, with some windings, through the great Bear onwards to Cassiopeia; thence through the girdle of Andromeda and the northern Fish, proceeding towards Cetus; while towards the south it passes through the Virgin, probably on to the tail of Hydra and the head of Centaurus. But, notwithstanding I have already fully ascertained the existence and direction of this stratum for more than 30 degrees of a great

[12] [See figure on p. 51.] [13] [M44. – S.]

[14] [The stratum of Coma Berenices is very different. Described for the greater part correctly by Herschel, it is an elongated area of sky running up through Virgo and Coma into Ursa Major (though not beyond), remarkably rich in galaxies. It contains the Virgo cluster of galaxies and has been called the 'supergalaxy': Herschel was the first to recognize it. "the Coma itself", i.e. the loose cluster of stars that form the constellation of Coma Berenices, is of course a fortuitous foreground object.

(Note the confusion of thought here: Herschel has not clearly distinguished in his mind between strata of stars (the Galaxy), star clusters, and galaxies. In modern terms the stratum of Coma Berenices is an elongated association of galaxies, the Virgo Cluster. Herschel thought of it as being a stratum of stars mostly grouped in clusters, of which only Coma Berenices was near enough to be resolved. He probably had a similar picture of the "stratum of Cancer".) – D.]

[15] [See figure on p. 52.]

circle, and found it almost every where equally rich in fine nebulae, it still might be dangerous to proceed in more extensive conjectures, that have as yet no more than a precarious foundation. I shall therefore wait till the observations in which I am at present engaged shall furnish me with proper materials for the disquisition of so new a subject. And though my single endeavours should not succeed in a work that seems to require the joint effort of every astronomer, yet so much we may venture to hope, that, by applying ourselves with all our powers to the improvement of telescopes, which I look upon as yet in their infant state, and turning them with assiduity to the study of the heavens, we shall in time obtain some faint knowledge of, and perhaps be able partly to delineate, *the Interior Construction of the Universe*.

WILLIAM HERSCHEL.

Datchet near Windsor,
April, 1784.

1785: "On the construction of the heavens"

Summary: "... I begin with the formation of nebulae, which I divide into four forms. Then I give a table of Star-Gages; and I show that by this method of counting stars that are visible at one time in the field of view of one diameter I arrive at our approximate place in the universe. I show that it seems very likely that our star system is a detached nebula of the third form. I give a section of this great nebula from which one sees that it is quite limited at the sides but that it is far extended in length and breadth. The principles on which I base this argument are based on observation aided by calculation. I have already seen 1015 nebulae, among which are some that are also of the third form and which may be not only equal to but even greater than the nebula of which our Sun is one of the stars composing it. I also give the locations of some of these great nebulae. Among the curiosities of the heavens one ought to reckon one composed of a kind of ring of stars; it is M57. At the close of the paper I give the location of half-a-dozen planetary nebulae, as I call them. These are celestial bodies of which as yet we have no clear idea and which are perhaps of a type quite different from those that we are familiar with in the heavens. I have already found four that have a visible diameter of between 15 and 30 seconds. These bodies appear to have a disk that is rather like a planet, that is to say, of equal brightness all over, round or somewhat oval, and about as well-defined in outline as the disk of the planets, of a light strong enough to be visible with an ordinary telescope of only one foot, yet they have only the appearance of a star of about ninth magnitude."

Herschel to Jerôme Lalande, 17 March 1785
(in French).[1]

[1] [RAS W.1/1, 128–129.]

On the Construction of the Heavens[2]

Read February 3, 1785.

THE subject of the Construction of the Heavens, on which I have so lately ventured to deliver my thoughts to this Society, is of so extensive and important a nature, that we cannot exert too much attention in our endeavours to throw all possible light upon it; I shall, therefore, now attempt to pursue the delineations of which a faint outline was begun in my former paper.

By continuing to observe the heavens with my last constructed, and since that time much improved instrument, I am now enabled to bring more confirmation to several parts that were before but weakly supported, and also to offer a few still further extended hints, such as they present themselves to my present view. But first let me mention that, if we would hope to make any progress in an investigation of this delicate nature, we ought to avoid two opposite extremes, of which I can hardly say which is the most dangerous. If we indulge a fanciful imagination and build worlds of our own, we must not wonder at our going wide from the path of truth and nature; but these will vanish like the Cartesian vortices, that soon gave way when better theories were offered. On the other hand, if we add observation to observation, without attempting to draw not only certain conclusions, but also conjectural views from them, we offend against the very end for which only observations ought to be made. I will endeavour to keep a proper medium; but if I should deviate from that, I could wish not to fall into the latter error.

That the milky way is a most extensive stratum of stars of various sizes admits no longer of the least doubt; and that our sun is actually one of the heavenly bodies belonging to it is as evident. I have now viewed and gaged this shining zone in almost every direction, and find it composed of stars whose number, by the account of these gages, constantly increases and decreases in proportion to its apparent brightness to the naked eye. But in order to develop the ideas of the universe, that have been suggested by my late observations, it will be best to take the subject from a point of view at a considerable distance both of space and of time.

Theoretical view

Let us then suppose numberless stars of various sizes, scattered over an indefinite portion of space in such a manner as to be almost equally distributed throughout the whole. The laws of attraction, which no doubt extend to the

[2] [*Phil. Trans.*, lxxv (1785), 213–266.]

remotest regions of the fixed stars, will operate in such a manner as most probably to produce the following remarkable effects.

Formation of nebulae

Form I.[3] In the first place, since we have supposed the stars to be of various sizes, it will frequently happen that a star, being considerably larger than its neighbouring ones, will attract them more than they will be attracted by others that are immediately around them; by which means they will be, in time, as it were, condensed about a center; or, in other words, form themselves into a cluster of stars of almost a globular figure, more or less regularly so, according to the size and original distance of the surrounding stars. The perturbations of these mutual attractions must undoubtedly be very intricate, as we may easily comprehend by considering what Sir Isaac Newton says in the first book of his Principia, in the 38th and following problems; but in order to apply this great author's reasoning of bodies moving in ellipses to such as are here, for a while, supposed to have no other motion than what their mutual gravity has imparted to them, we must suppose the conjugate axes of these ellipses indefinitely diminished, whereby the ellipses will become straight lines.

Form II. The next case, which will also happen almost as frequently as the former, is where a few stars, though not superior in size to the rest, may chance to be rather nearer each other than the surrounding ones; for here also will be formed a prevailing attraction in the combined center of gravity of them all, which will occasion the neighbouring stars to draw together; not indeed so as to form a regular or globular figure, but however in such a manner as to be condensed towards the common center of gravity of the whole irregular cluster. And this construction admits of the utmost variety of shapes, according to the number and situation of the stars which first gave rise to the condensation of the rest.

Form III. From the composition and repeated conjunction of both the foregoing forms, a third may be derived, when many large stars, or combined small ones, are situated in long extended, regular, or crooked rows, hooks, or branches; for they will also draw the surrounding ones, so as to produce figures of condensed stars coarsely similar to the former which gave rise to these condensations.

Form IV. We may likewise admit of still more extensive combinations; when, at the same time that a cluster of stars is forming in one part of space, there may

[3] [These four *Forms* of nebulae, referred to by Herschel in papers about this time, are not to be confused with the eight *Classes* of nebulae and clusters into which Herschel divided the objects in his catalogues on observational grounds. – D.]

be another collecting in a different, but perhaps not far distant quarter, which may occasion a mutual approach towards their common center of gravity.

[Form] V. In the last place, as a natural consequence of the former cases, there will be formed great cavities or vacancies by the retreat of the stars towards the various centers which attract them; so that upon the whole there is evidently a field of the greatest variety for the mutual and combined attractions of the heavenly bodies to exert themselves in. I shall, therefore, without extending myself farther upon this subject, proceed to a few considerations, that will naturally occur to every one who may view this subject in the light I have here done.

Objections considered

At first sight then it will seem as if a system, such as it has been displayed in the foregoing paragraphs, would evidently tend to a general destruction, by the shock of one star's falling upon another. It would here be a sufficient answer to say, that if observation should prove this really to be the system of the universe, there is no doubt but that the great Author of it has amply provided for the preservation of the whole, though it should not appear to us in what manner this is effected. But I shall moreover point out several circumstances that do manifestly tend to a general preservation; as, in the first place, the indefinite extent of the sidereal heavens, which must produce a balance that will effectually secure all the great parts of the whole from approaching to each other. There remains then only to see how the particular stars belonging to separate clusters will be preserved from rushing on to their centers of attraction. And here I must observe, that though I have before, by way of rendering the case more simple, considered the stars as being originally at rest, I intended not to exclude projectile forces; and the admission of them will prove such a barrier against the seeming destructive power of attraction as to secure from it all the stars belonging to a cluster, if not for ever, at least for millions of ages. Besides, we ought perhaps to look upon such clusters, and the destruction of now and then a star, in some thousands of ages, as perhaps the very means by which the whole is preserved and renewed. These clusters may be the *Laboratories* of the universe, if I may so express myself, wherein the most salutary remedies for the decay of the whole are prepared.

Optical appearances

From this theoretical view of the heavens, which has been taken, as we observed, from a point not less distant in time than in space, we will now retreat to our own retired station, in one of the planets attending a star in its great combination with numberless others;[4] and in order to investigate what will be the appearances from this contracted situation, let us begin with the naked eye. The stars of the first magnitude being in all probability the nearest, will furnish us with a step to begin our scale; setting off, therefore, with the distance of Sirius or Arcturus, for instance, as unity, we will at present suppose, that those of the second magnitude are at double, and those of the third at treble the distance, and so forth. It is not necessary critically to examine what quantity of light or magnitude of a star intitles it to be estimated of such or such a proportional distance, as the common coarse estimation will answer our present purpose as well; taking it then for granted, that a star of the seventh magnitude is about seven times as far as one of the first, it follows, that an observer, who is inclosed in a globular cluster of stars, and not far from the center, will never be able, with the naked eye, to see to the end of it: for, since, according to the above estimations, he can only extend his view to about seven times the distance of Sirius, it cannot be expected that his eyes should reach the borders of a cluster which has perhaps not less than fifty stars in depth every where around him. The whole universe, therefore, to him will be comprised in a set of constellations, richly ornamented with scattered stars of all sizes. Or if the united brightness of a neighbouring cluster of stars should, in a remarkable clear night, reach his sight, it will put on the appearance of a small, faint, whitish, nebulous cloud, not to be perceived without the greatest attention. To pass by other situations, let him be placed in a much extended stratum, or branching cluster of millions of stars, such as may fall under the IIId form of nebula considered in a foregoing paragraph. Here also the heavens will not only be richly scattered over with brilliant constellations, but a shining zone or milky way will be perceived to surround the whole sphere of the heavens, owing to the combined light of those stars which are too small, that is, too remote to be seen. Our observer's sight will be so confined, that he will imagine this single collection of stars, of which he does not even perceive the thousandth part, to be the whole contents of the heavens. Allowing him now the use of a common telescope, he begins to suspect that all the milkiness of the bright path which surrounds the sphere may be owing to stars. He perceives a few clusters of them in various parts of the heavens, and finds also that there are a kind of nebulous patches; but still his views are not extended so far as to reach to the end of the stratum in which he is situated, so that he looks upon these patches as belonging to that system which to him seems to comprehend every celestial object. He now increases his power of vision, and, applying himself to a close observation, finds that the milky way

[4] [In Herschel's manuscript, heavily deleted but just legible, are the words: "that make up a cluster, which itself is but one of the minute parts of an indefinite whole".]

is indeed no other than a collection of very small stars. He perceives that those objects which had been called nebulae are evidently nothing but clusters of stars. He finds their number increase upon him, and when he resolves one nebula into stars he discovers ten new ones which he cannot resolve. He then forms the idea of immense strata of fixed stars, of clusters of stars and of nebulae;[5] till, going on with such interesting observations, he now perceives that all these appearances must naturally arise from the confined situation in which we are placed. *Confined* it may justly be called, though in no less a space than what before appeared to be the whole region of the fixed stars; but which now has assumed the shape of a crookedly branching nebula; not, indeed, one of the least, but perhaps very far from being the most considerable of those numberless clusters that enter into the construction of the heavens.

Result of Observations

I shall now endeavour to shew, that the theoretical view of the system of the universe, which has been exposed in the foregoing part of this paper, is perfectly consistent with facts, and seems to be confirmed and established by a series of observations. It will appear, that many hundreds of nebulae of the first and second forms are actually to be seen in the heavens, and their places will hereafter be pointed out. Many of the third form will be described, and instances of the fourth related. A few of the cavities mentioned in the fifth will be particularised, though many more have already been observed; so that, upon the whole, I believe, it will be found, that the foregoing theoretical view, with all its consequential appearances, as seen by an eye inclosed in one of the nebulae, is no other than a drawing from nature, wherein the features of the original have been closely copied; and I hope the resemblance will not be called a bad one, when it shall be considered how very limited must be the pencil of an inhabitant of so small and retired a portion of an indefinite system in attempting the picture of so unbounded an extent.

But to proceed to particulars: I shall begin by giving the following table of gages that have been taken. In the first column is the right ascension, and in the second the north polar distance, both reduced to the time of FLAMSTEED's Catalogue. In the third are the contents of the heavens, being the result of the gages. The fourth shews from how many fields of view the gages were deduced, which have been ten or more where the number of the stars was not very considerable; but, as it would have taken too much time, in high numbers, to count so many fields, the gages are generally single. Where the stars happened to be uncommonly crowded, no more than half a field

[5] See a former paper on the Construction of the Heavens.

was counted, and even sometimes only a quadrant; but then it was always done with the precaution of fixing on some row of stars that would point out the division of the field, so as to prevent any considerable mistake. When five, ten, or more fields are gaged, the polar distance in the second column of the table is that of the middle of the sweep, which was generally from 2 to 2½ degrees in breadth; and, in gaging, a regular distribution of the fields, from the bottom of the sweep to the top, was always strictly attended to. The fifth column contains occasional remarks relating to the gages.

I. *Table of Star-Gages.*

R.A.			P.D.		Stars.	Fields.	Memorandum
H.	M.	S.	D.	M.			
0	1	41	78	47	9.9	10	
0	4	55	65	36	20.0	10	
0	7	54	74	13	11.3	10	Most of the stars extremely small.
0	8	24	49	7	60	1	
0	9	52	113	17	4.1	10	
23	46	55	65	36	15.3	10	
23	59	21	87	10	5.6	10	
23	59	56	95	4	7.8	10	

[135 sets of five gages each are not reproduced.]

PROBLEM

The stars being supposed to be nearly equally scattered, and their number, in a field of view of a known angular diameter, being given, to determine the length of the visual ray.

[Calculations not reproduced.]

We inhabit the planet of a star belonging to a Compound Nebula of the third form.

I shall now proceed to shew that the stupendous sidereal system we inhabit, this extensive stratum and its secondary branch, consisting of many millions of stars, is, in all probability, *a detached Nebula*. In order to go upon grounds that seem to me to be capable of great certainty, they being no less than an actual survey of the boundaries of our sidereal system, which I have plainly perceived, as far as I have yet gone round it, every where terminated, and in most places very narrowly too, it will be proper to shew the length of my sounding line, if I may so call it, that it may appear whether it was sufficiently long for the purpose.

In the most crowded part of the milky way I have had fields of view that contained no less than 588 stars,[6] and these were continued for many minutes, so that in one quarter of an hour's time there passed no less than 116000 stars through the field of view of my telescope.[7] Now, if we compute the length of the visual ray by putting S = 588, and the diameter of the field of view fifteen minutes, we shall find $n = \sqrt[3]{(B^2S)} = 498$; so that it appears the length of what I have called my sounding line, or $n - 1$, was probably not less than 497 times the distance of Sirius from the sun. . . .

It may seem inaccurate that we should found an argument on the stars being equally scattered, when in all probability there may not be two of them in the heavens, whose mutual distance shall be equal to that of any other two given stars; but it should be considered, that when we take all the stars collectively there will be a mean distance which may be assumed as the general one; and an argument founded on such a supposition will have in its favour the greatest probability of not being far short of truth. What will render the supposition of an equal distribution of the stars, with regard to the gages, still less exposed to objections is, that whenever the stars happened either to be uncommonly crowded or deficient in number, so as very suddenly to pass over from one extreme to the other, the gages were reduced to other forms, such as the border-gage, the distance-gage, &c. which terms, and the use of such gages, I shall hereafter find an opportunity of explaining. And none of those kinds of gages have been admitted in this table, which consists only of such as have been taken in places where the stars apparently seemed to be, in general, pretty evenly scattered; and to increase and decrease in number by a certain gradual progression. Nor has any part of the heavens containing a cluster of stars been put in the gages; and here I must observe, that the difference between a crowded place and a cluster may easily be perceived by the arrangement as well as the size and mutual distance of the stars: for in a cluster they are generally not only resembling each other pretty nearly in size, but a certain uniformity of distance also takes place; they are more and more accumulated towards the center, and put on all the appearances which we should naturally expect from a number of them collected into a group at a certain distance from us. On the other hand, the rich parts of the milky way, as well as those in the distant broad part of the stratum, consist of a mixture of stars of all possible sizes, that are seemingly

[6] See the table of Gages.

[7] The breadth of my sweep was 2° 26′, to which must be added 15′ for two semi-diameters of the field. Then putting 161 = a, the number of fields in 15 minutes of time; ,7854 = b, the proportion of a circle to 1, its circumscribed square; φ = sine of 74° 22′, the polar distance of the middle of the sweep reduced to the present time; and 588 = S, the number of stars in a field of view, we have $a\varphi S/b$ = 116076 stars.

placed without any particular apparent order. Perhaps we might recollect, that a greater condensation towards the center of our system than towards the borders of it should be taken into consideration; but, with a nebula of the third form, containing such various and extensive combinations, as I have found to take place in ours, this circumstance, which in one of the first form would be of considerable moment, may, I think, be safely neglected. However, I would not be understood to lay a greater stress on these and the following calculations than the principles on which they are founded will permit; and if hereafter we shall find reason, from experience and observation, to believe that there are parts of our system where the stars are not scattered in the manner here supposed, we ought then to make proper exceptions.

But to return: if some other high gage be selected from the table, such as 472 or 344, the length of the visual ray will be found 461 and 415. And although, in consequence of what has been said, a certain degree of doubt may be left about the arrangement and scattering of the stars, yet when I recollect, that in those parts of the milky way where these high gages were taken, the stars were neither so small, nor so crowded, as they must have been on a supposition of a much farther continuance of them, when certainly a milky or nebulous appearance must have come on, I need not fear to have over-rated the extent of my visual ray. And indeed every thing that can be said to shorten it will only contract the limits of our nebula, as it has in most places been of sufficient length to go far beyond the bounds of it. Thus, in the sides of the stratum opposite to our situation in it, where the gages often run below 5, our nebula cannot extend to 100 times the distance of Sirius; and the same telescope, which could shew 588 stars in a field of view of 15 minutes, must certainly have presented me also with the stars in these situations as well as the former, had they been there. If we should answer this by observing that they might be at too great a distance to be perceived, it will be allowing that there must at least be a vacancy amounting to the length of a visual ray not short of 400 times the distance of Sirius; and this is amply sufficient to make our nebula a detached one. It is true, that it would not be consistent confidently to affirm that we were on an island unless we had actually found ourselves every where bounded by the ocean, and therefore I shall go no farther than the gages will authorise; but considering the little depth of the stratum in all those places which have been actually gaged, to which must be added all the intermediate parts that have been viewed and found to be much like the rest, there is but little room to expect a connection between our nebula and any of the neighbouring ones. I ought also to add, that a telescope with a much larger aperture than my present one, grasping together a greater quantity of light, and thereby enabling us to see farther into space, will be the surest means of compleating and establishing the arguments that have been used: for if our nebula is not absolutely a detached one, I am firmly persuaded, that an instrument may be made large enough to discover the places where the stars continue onwards. A very bright milky nebulosity must there undoubtedly come on, since the stars in a field of view will increase in the ratio of n^3, greater than that of the cube of the visual ray. Thus, if 588 stars in a given field of view are to be seen by a ray of 497 times the distance of Sirius; when this is lengthened to 1000, which is but little more than double the former, the number of stars in the same field of view will be no less than 4774: for when the visual ray r is given, the number S of stars will be $= n^3/B^2$; where $n = r + 1$; and a telescope with a three-fold power of extending into space, or with a ray of 1500, which, I think, may easily be constructed, will give us 16096 stars. Now, these would not be so close but that a good power applied to such an instrument might easily distinguish them; for they need not, if arranged in regular squares, approach nearer to each other than 6",27; but what would produce the milky nebulosity which I have mentioned is the numberless stars beyond them, which in one respect the visual ray might also be said to reach. To make this appear we must return to the naked eye, which, as we have before estimated, can only see the stars of the seventh magnitude so as to distinguish them; but it is nevertheless very evident that the united lustre of millions of stars, such as I suppose the nebula in Andromeda to be, will reach our sight in the shape of a very small, faint nebulosity; since the nebula of which I speak may easily be seen in a fine evening. In the same manner my present telescope, as I have argued, has not only a visual ray that will reach the stars at 497 times the distance of Sirius so as to distinguish them (and probably much farther), but also a power of shewing the united lustre of the accumulated stars that compose a milky nebulosity, at a distance far exceeding the former limits; so that from these considerations it appears again highly probable, that my present telescope, not shewing such a nebulosity in the milky way, goes already far beyond its extent: and consequently, much more would an instrument, such as I have mentioned, remove all doubt on the subject, both by shewing the stars in the continuation of the stratum, and by exposing a very strong milky nebulosity beyond them, that could no longer be mistaken for the dark ground of the heavens.

To these arguments, which rest on the firm basis of a series of observations, we may add the following considerations drawn from analogy. Among the great number of nebulae which I have now already seen, amounting to more than 900, there are many which in all probability are equally extensive with that which we inhabit; and yet they are all separated from each other by very considerable intervals. Some indeed there are that seem to be double and treble; and though with most of these it may be, that they are at a very great distance from each other, yet we allow that some such conjunctions really are to be found; nor is this what we mean to exclude. But then these compound or double nebulae, which are those of the third and fourth forms, still make a detached link in the great chain. It is also to be supposed, that there may still be some thinly scattered solitary stars between the large interstices of nebulae, which, being situated so as to be nearly equally attracted by the several clusters when they were forming, remain unassociated. And though we cannot expect to see these stars, on account of their vast distance, yet we may well presume, that their number cannot be very considerable in comparison to those that are already drawn into systems; which conjecture is also abundantly confirmed in situations where the nebulae are near enough to have their stars visible; for they are all insulated, and generally to be seen upon a very clear and pure ground, without any star near them that might be supposed to belong to them. And though I have often seen them in beds of stars, yet from the size of these latter we may be certain, that they were much nearer to us than those nebulae, and belonged undoubtedly to our own system.

Use of the gages

A delineation of our nebula, by an application of the gages in the manner which has been proposed to be done in my former paper, may now be attempted, and the following table [Table II] is calculated for this purpose. It gives us the length of the visual ray for any number of stars in the field of view contained in the third column of the foregoing table of gages from 1/10 to 100000. If the number required is not to be found in the first column of this table, a proportional mean may be taken between the two nearest rays in the second column, without any material error, except in the few last numbers. The calculations of resolvable and milky nebulosity, at the end of the table, are founded, the first, on a supposition of the stars being so crowded as to have only a square second of space allowed them; the next assigning them only half a second square. However, we should consider that in all probability a very different accumulation of stars may take place in different nebulae; by which means some of them may assume the milky appearance, though not near so far removed from us; while clusters of stars also may become resolvable nebulae from the same cause. The distinctness of the instrument is here also concerned; and as telescopes with large apertures are not easily brought to a good figure, nebulous appearances of both sorts may probably come on much before the distance annexed to them in the table.

Table II.

Stars in the field.	Visual ray.	Stars.	Ray.	Stars.	Ray.	Stars.	Ray.	Stars.	Ray.
		31	186	71	245	210	352	700	527
0,1	27	32	188	72	246	220	358	800	551
0,2	34	33	190	73	247	230	363	900	573
0,3	39	34	192	74	249	240	368	1000	593
0,4	43	35	193	75	250	250	374	10000	1280
0,5	46	36	195	76	251	260	378	100000	2758
0,6	49	37	197	77	252	270	383		
0,7	52	38	199	78	252	280	388		
0,8	54	39	201	79	254	290	393		
0,9	56	40	202	80	255	300	397		
1	58	41	204	81	256	310	401		
2	74	42	206	82	257	320	406	636175 or resolvable nebulosity	5112
3	85	43	207	83	258	330	410		
4	93	44	209	84	259	340	414		
5	101	45	210	85	260	350	418		
6	107	46	212	86	261	360	422		
7	113	47	214	87	262	370	426		
8	118	48	215	88	263	380	430		
9	123	49	217	89	264	390	433		
10	127	50	218	90	265	400	437		
11	131	51	219	91	266	410	441		
12	135	52	221	92	267	420	444	2544700 or milky nebulosity	8115
13	139	53	222	93	268	430	448		
14	142	54	224	94	269	440	451		
15	146	55	225	95	270	450	455		
16	149	56	226	96	271	460	458		
17	152	57	228	97	272	470	461		
18	155	58	229	98	273	480	464		
19	158	59	230	99	274	490	468		
20	160	60	232	100	275	500	471		
21	163	61	233	110	284	510	474		
22	166	62	234	120	291	520	477		
23	168	63	236	130	300	530	480		
24	170	64	237	140	308	540	483		
25	173	65	238	150	315	550	486		
26	175	66	239	160	322	560	489		
27	177	67	240	170	328	570	492		
28	180	68	242	180	335	580	495		
29	182	69	243	190	341	590	498		
30	184	70	244	200	347	600	500		

Section of our sidereal system

By taking out of this table the visual rays which answer to the gages, and applying lines proportional to them around a point, according to their respective right ascensions and north polar distances, we may delineate a solid by means of the ends of these lines, which will give us so many points in its surface; I shall, however, content myself at present with a section only. I have taken one which passes through the poles of our system, and is at rectangles to the conjunction of the branches which I have called its length. The name of poles seemed to me not improperly applied to those points which are 90 degrees distant from a circle passing along the milky way, and the north pole is here assumed to be situated in R.A. 186° and P.D. 58°. The section represented in fig. 4 is one which makes an angle of 35 degrees with our equator, crossing it in 124½ and 304½ degrees. A celestial globe, adjusted to the latitude of 55° north, and having σ Ceti near the meridian, will have the plane of this section pointed out by the horizon, and the gages which have been used in this delineation are those which in table I. are marked by asterisks. When the visual rays answering to them are taken out of the second table, they must be projected on the plane of the horizon of the latitude which has been pointed out; and this may be done accurately enough for the present purpose by a globe adjusted as above directed; for as gages, exactly in the plane of the section, were often wanting, I have used many at some small distance above and below the same, for the sake of obtaining more delineating points; and in the figure the stars at the borders which are larger than the rest are those pointed out by the gages. The intermediate parts are filled up by smaller stars arranged in straight lines between the gaged ones. The delineating points, though pretty numerous, are not so close as we might wish; it is however to be hoped that in some future time this branch of astronomy will become more cultivated, so that we may have gages for every quarter of a degree of the heavens at least, and these often repeated in the most favourable circumstances. And whenever that shall be the case, delineations may then be repeated with all the accuracy that long experience may enable us to introduce; for, this subject being so new, I look upon what is here given partly as only an example to illustrate the spirit of the method. From this figure, however, which I hope is not a very inaccurate one, we may see that our nebula, as we observed before, is of the third form; that is: *A very extensive, branching, compound Congeries of many millions of stars*; which most probably owes its origin to many remarkably large as well as pretty closely scattered small stars, that may have drawn together the rest. Now, to have some idea of the wonderful extent of this system, I must observe that this section of it is drawn upon a scale where the distance of Sirius is no more than the [135th] part of an inch; so that probably all the stars,

Fig. 4.

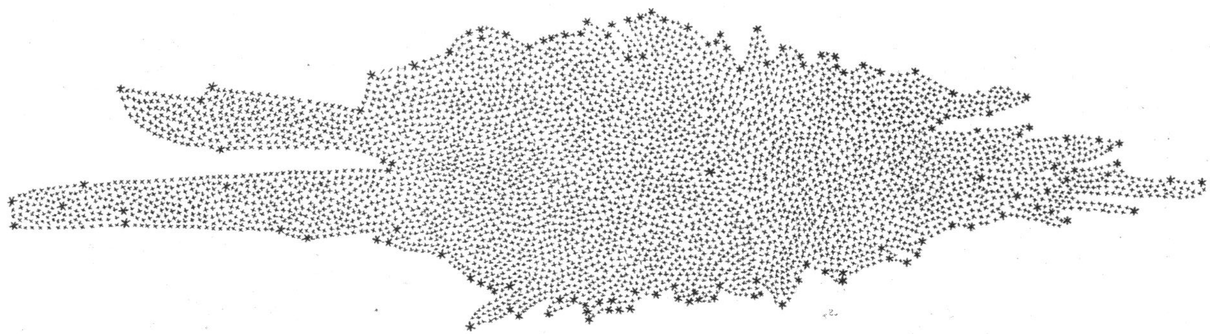

Fig. 5.

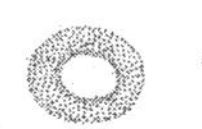

which in the finest nights we are able to distinguish with the naked eye, may be comprehended within a sphere, drawn round the large star near the middle, representing our situation in the nebula, of less than half a quarter of an inch radius.

The Origin of nebulous Strata

If it were possible to distinguish between the parts of an indefinitely extended whole, the nebula we inhabit might be said to be one that has fewer marks of profound antiquity upon it than the rest. To explain this idea perhaps more clearly, we should recollect that the condensation of clusters of stars has been ascribed to a gradual approach; and whoever reflects on the numbers of ages that must have past before some of the clusters, that will be found in my intended catalogue of them, could be so far condensed as we find them at present, will not wonder if I ascribe a certain air of youth and vigour to many very regularly scattered regions of our sidereal stratum. There are moreover many places in it where there is the greatest reason to believe that the stars, if we may judge from appearances, are now drawing towards various secondary centers, and will in time separate into different clusters, so as to occasion many sub-divisions. Hence we may surmise that when a nebulous stratum consists chiefly of nebulae of the first and second form, it probably owes its origin to what may be called the decay of a great compound nebula of the third form; and that the sub-divisions, which happened to it in length of time, occasioned all the small nebulae which sprung from it to lie in a certain range, according as they were detached from the primary one. In like manner our system, after numbers of ages, may very possibly become divided so as to give rise to a stratum of two or three hundred nebulae; for it would not be difficult to point out so many beginning or gathering clusters in it.[8] This view of the present subject throws a considerable light upon the appearance of that remarkable collection of many hundreds of nebulae which are to be seen in what I have called the nebulous stratum of Coma Berenices. It appears from the extended and branching figure of our nebula, that there is room for the decomposed small nebula of a large, reduced, former great one to approach nearer to us in the sides than in other parts. Nay, possibly, there might originally be another very large joining branch, which in time became separated by the condensation of the stars; and this may

[8] Mr. MICHELL has also considered the stars as gathered together into groups (Phil. Trans. vol. LVII. p. 249); which idea agrees with the sub-division of our great system here pointed out. He founds an elegant proof of this on the computation of probabilities, and mentions the Pleiades, the Praesepe Cancri, and the nebula (or cluster of stars) in the hilt of Perseus's sword, as instances. [The last-named object is the double-cluster χ Persei, NGC 869/884. – S.]

be the reason of the little remaining breadth of our system in that very place: for the nebulae of the stratum of the Coma are brightest and most crowded just opposite our situation, or in the pole of our system. As soon as this idea was suggested, I tried also the opposite pole, where accordingly I have met with a great number of nebulae, though under a much more scattered form.

An Opening in the heavens

Some parts of our system indeed seem already to have sustained greater ravages of time than others, if this way of expressing myself may be allowed; for instance, in the body of the Scorpion is an opening, or hole, which is probably owing to this cause. I found it while I was gaging in the parallel from 112 to 114 degrees of north polar distance. As I approached the milky way, the gages had been gradually running up from 9,7 to 17,1; when, all of a sudden, they fell down to nothing, a very few pretty large stars excepted, which made them shew 0,5, 0,7, 1,1, 1,4, 1,8; after which they again rose to 4,7, 13,5, 20,3, and soon after to 41,1. This opening is at least 4 degrees broad, but its height I have not yet ascertained. It is remarkable, that the 80th *Nébuleuse sans étoiles* of the *Connoissance des Temps,* which is one of the richest and most compressed clusters of small stars I remember to have seen, is situated just on the western border of it, and would almost authorise a suspicion that the stars, of which it is composed, were collected from that place, and had left the vacancy. What adds not a little to this surmise is, that the same phaenomenon is once more repeated with the fourth cluster of stars of the *Connoissance des Temps*; which is also on the western border of another vacancy, and has moreover a small, miniature cluster, or easily resolvable nebula of about 2½ minutes in diameter, north following it, at no very great distance.[9]

Phaenomena at the Poles of our Nebula

I ought to observe, that there is a remarkable purity or clearness in the heavens[10] when we look out of our stratum at the sides; that is, towards Leo, Virgo, and Coma Berenices, on one hand, and towards Cetus on the other; whereas the ground of the heavens becomes troubled as we approach towards the length or height of it. It was a good while before I could trace the cause of these phaenomena; but since I have been acquainted with the shape of our system, it is plain that these troubled appearances, when we approach to the sides, are easily

[9] [The star clusters here referred to are the globular clusters M80, M4, and NGC 6144. They lie in the vicinity of an area of heavy obscuration that cuts out the light from more distant stars. The contribution of interstellar dust to the patchy distribution of stars in the Milky Way has been fully recognized only in recent decades. – D.]

[10] [Herschel correctly interprets an observation familiar to visual observers. – D.]

to be explained by ascribing them to some of the distant, straggling stars, that yield hardly light enough to be distinguished. And I have, indeed, often experienced this to be actually the cause, by examining these troubled spots for a long while together, when, at last, I generally perceived the stars which occasioned them. But when we look towards the poles of our system, where the visual ray does not graze along the side, the straggling stars of course will be very few in number; and therefore the ground of the heavens will assume that purity which I have always observed to take place in those regions.

Enumeration of very compound Nebulae or Milky-Ways

As we are used to call the appearance of the heavens, where it is surrounded with a bright zone, the Milky-Way, it may not be amiss to point out some other very remarkable Nebulae which cannot well be less, but are probably much larger than our own system; and, being also extended, the inhabitants of the planets that attend the stars which compose them must likewise perceive the same phaenomena. For which reason they may also be called milky-ways by way of distinction.

My opinion of their size is grounded on the following observations. There are many round nebulae, of the first form, of about five or six minutes in diameter, the stars of which I can see very distinctly; and on comparing them with the visual ray calculated from some of my long gages, I suppose, by the appearance of the small stars in those gages, that the centers of these round nebula may be 600 times the distance of Sirius from us.

In estimating the distance of such clusters I consulted rather the comparatively apparent size of the stars than their mutual distance; for the condensation in these clusters being probably much greater than in our own system, if we were to overlook this circumstance and calculate by their apparent compression, where, in about six minutes diameter, there are perhaps ten or more stars in the line of measures, we should find, that on the supposition of an equal scattering of the stars throughout all nebulae, the distance of the center of such a cluster from us could not be less than 6000 times the distance of Sirius. And, perhaps, in putting it, by the apparent size of the stars, at 600 only, I may have considerably underrated it; but my argument, if that should be the case, will be so much the stronger. Now to proceed.

Some of these round nebulae have others near them, perfectly similar in form, colour, and the distribution of stars, but of only half the diameter: and the stars in them seem to be doubly crowded, and only at about half the distance from each other: they are indeed so small as not to be visible without the utmost attention. I suppose these miniature nebulae to be at double the distance of the first. An instance, equally remarkable and instructive, is a case where, in the neighbourhood of two such nebulae as have been mentioned, I met with a third, similar, resolvable, but much smaller and fainter nebula. The stars of it are no longer to be perceived; but a resemblance of colour with the former two, and its diminished size and light, may well permit us to place it at full twice the distance of the second, or about four or five times that of the first. And yet the nebulosity is not of the milky kind; nor is it so much as difficultly resolvable, or colourless. Now, in a few of the extended nebulae, the light changes gradually so as from the resolvable to approach to the milky kind; which appears to me an indication that the milky light of nebulae is owing to their much greater distance. A nebula, therefore, whose light is perfectly milky, cannot well be supposed to be at less than six or eight thousand times the distance of Sirius; and though the numbers here assumed are not to be taken otherwise than as very coarse estimates, yet an extended nebula, which in an oblique situation, where it is possibly fore-shortened by one-half, two-thirds, or three-fourths of its length, subtends a degree or more in diameter, cannot be otherwise than of a wonderful magnitude, and may well outvie our milky-way in grandeur.

The first I shall mention is a milky Ray of more than a degree in length. It takes *k* (Fl. 52.) Cygni into its extent, to the north of which it is crookedly bent so as to be convex towards the following side; and the light of it is pretty intense. To the south of *k* it is more diffused, less bright, and loses itself with some extension in two branches, I believe; but for want of light I could not determine this circumstance. The northern half is near two minutes broad, but the southern is not sufficiently defined to ascertain its breadth.

The next is an extremely faint milky Ray, above ¾ degree long, and 8 or 10′ broad; extended from north preceding to south following. It makes an angle of about 30 or 40 degrees with the meridian, and contains three or four places that are brighter than the rest. The stars of the Galaxy are scattered over it in the same manner as over the rest of the heavens. It follows ε Cygni 11,5 minutes in time, and is 2° 19′ more south.

The third is a branching Nebulosity of about a degree and a half in right ascension, and about 48′ extent in polar distance. The following part of it is divided into several streams and windings, which, after separating, meet each other again towards the south. It precedes ζ Cygni 16′ in time, and is 1° 16′ more north. I suppose this to be joined to the preceding one; but having observed them in different sweeps, there was no opportunity of tracing their connection.[11]

[11] ["a milky Ray" (= NGC 6960), "an extremely faint milky Ray", "a branching Nebulosity". These last two are part of the same object as Herschel surmises (NGC 6992). The three constitute peripheral arcs of a more or less circular area of slowly expanding emission

The fourth is a faint, extended milky Ray of about 17′ in length, and 12′ in breadth.[12] It is brightest and broadest in the middle, and the ends lose themselves. It has a small, round, very faint nebula just north of it; and also, in another place, a spot, brighter than the rest, almost detached enough to form a different nebula, but probably belonging to the great one. The Ray precedes α Trianguli 18′,8 in time, and is 55′ more north. Another observation of the same, in a finer evening, mentions its extending much farther towards the south, and that the breadth of it probably is not less than half a degree; but being shaded away by imperceptible gradations, it is difficult exactly to assign its limits.

The fifth is a Streak of light about 27′ long, and in the brightest part 3 or 4′ broad. The extent is nearly in the meridian, or a little from south preceding to north following. It follows β Ceti 5′,9 in time, and is 2° 43′ more south. The situation is so low, that it would probably appear of a much greater extent in a higher altitude.[13]

The sixth is an extensive milky Nebulosity divided into two parts; the most north being the strongest. Its extent exceeds 15′; the southern part is followed by a parcel of stars which I suppose to be the 8th of the *Connoissance des Temps*.[14]

The seventh is a wonderful, extensive Nebulosity of the milky kind. There are several stars visible in it, but they can have no connection with that nebulosity, and are, doubtless, belonging to our own system scattered before it. It is the 17th of the *Connoissance des Temps*.[15]

In the list of these must also be reckoned the beautiful Nebula of Orion.[16] Its extent is much above one degree; the eastern branch passes between two very small stars, and runs on till it meets a very bright one. Close to the four small stars, which can have no connection with the nebula, is a total blackness; and within the open part, towards the north-east, is a distinct, small, faint nebula,[17] of an extended shape, at a distance from the border of the great one, to which it runs in a parallel direction, resembling the shoals that are seen near the coasts of some islands.

nebulosity some two and a half degrees in diameter, which is thought to be the remnants of a prehistoric supernova (the 'Veil' nebula, 'Cygnus Loop'). – D.]

12 [NGC 598, which Herschel does not seem to have realized to be M33, to which he has referred earlier. The several patches of nebulosity Herschel refers to are the brighter parts of what is really one object, a near-by, face-on Sc galaxy some 45′ in overall diameter. – D.]

13 [NGC 247, elongated Sc galaxy. – D.]

14 [The 'Lagoon' nebula, a bright gaseous nebulosity associated with the star cluster M8 = NGC 6523/6533. – D.]

15 [M17 = NGC 6618. Gaseous nebulosity, the 'Horseshoe' or 'Omega' nebula. – D.]

16 [The Orion Nebula, M42 = NGC 1976: the finest example in the sky of gaseous nebulosity with associated stars and obscuring matter. – D.]

17 [M43. – S.]

The ninth is that in the girdle of Andromeda,[18] which is undoubtedly the nearest of all the great nebulae; its extent is above a degree and a half in length, and, in even one of the narrowest places, not less than 16′ in breadth. The brightest part of it approaches to the resolvable nebulosity, and begins to shew a faint red colour; which, from many observations on the colour and magnitude of nebulae, I believe to be an indication that its distance in this coloured part does not exceed 2000 times the distance of Sirius. There is a very considerable, broad, pretty faint, small nebula near it; my Sister discovered it August 27, 1783, with a Newtonian 2-feet sweeper. It shews the same faint colour with the great one, and is, no doubt, in the neighbourhood of it. It is not the 32d of the *Connoissance des Temps*; which is a pretty large round nebula, much condensed in the middle, and south following the great one; but this is about two-thirds of a degree north preceding it, in a line parallel to β and ν Andromedae.

To these may be added the nebula in Vulpecula:[19] for, though its appearance is not large, it is probably a double stratum of stars of a very great extent, one end whereof is turned towards us. That it is thus situated may be surmised from its containing, in different parts, nearly all the three nebulosities; *viz.* the resolvable, the coloured but irresolvable, and a tincture of the milky kind. Now, what great length must be required to produce these effects may easily be conceived when, in all probability, our whole system, of about 800 stars in diameter, if it were seen at such a distance that one end of it might assume the

[18] [The Andromeda Nebula, M31 = NGC 224. Herschel's interpretation of it as a near-by galaxy is of course correct, but he was wrong in supposing "approaching resolvability" and the distance is perhaps 300,000 times that of Sirius, not 2000. The nebula near it, discovered independently by Caroline Herschel in 1783 (but previously seen by Messier ten years earlier), is NGC 205, and M32 = NGC 221. Both are small elliptical galaxies close in space to the large spiral.

Herschel was a most experienced observer, and may be supposed to have been aware of the relative colours of the brighter nebula and the effect of his specula upon them. Some weight may be attached to his record that the brighter (i.e. nuclear) parts of M31, and the galaxy NGC 205, "shew a faint red colour". If accepted, it is one of Herschel's most remarkable observations, anticipating by a hundred and fifty years an observation that is still made visually with difficulty, but confirmed by photometry: the nuclear regions of spiral galaxies like M31 are redder than the outer parts because of the preponderance of red giant stars of Population II. The visual observation is usually attributed to Baade, using the 100-inch telescope on Mount Wilson in about 1942. – D.]

[19] [M27 = NGC 6853, a large planetary nebula. Although roughly circular in shape, the distribution of light gives it its popular name of the 'Dumbbell' and accounts for Herschel's interpretation of it as a double stratum of stars seen end-on.

Note that of Herschel's nine objects listed above as type examples of "Milky Ways" like our own, only three are in fact galaxies. – D.]

resolvable nebulosity, would not, at the other end, present us with the irresolvable, much less with the colourless and milky sort of nebulosities.

A Perforated Nebula, or Ring of Stars

Among the curiosities of the heavens should be placed a nebula, that has a regular, concentric, dark spot in the middle, and is probably a Ring of stars.[20] It is of an oval shape, the shorter axis being to the longer as about 83 to 100; so that, if the stars form a circle, its inclination to a line drawn from the sun to the center of this nebula must be about 56 degrees. The light is of the resolvable kind, and in the northern side three very faint stars may be seen, as also one or two in the southern part. The vertices of the longer axis seem less bright and not so well defined as the rest. There are several small stars very near, but none that seem to belong to it. It is the 57th of the *Connoissance des Temps*. Fig. 5 is a representation of it.

Planetary Nebulae

I shall conclude this paper with an account of a few heavenly bodies, that from their singular appearance leave me almost in doubt where to class them.[21]

The first precedes ν Aquarii 5′,4 in time, and is 1′ more north.[22] Its place, with regard to a small star Sept. 7, 1782, was, Distance 8′ 13″ 51‴; but on account of the low situation, and other unfavourable circumstances, the measure cannot be very exact. August 25, 1783, Distance 7′ 5″ 11‴, very exact, and to my satisfaction; the light being thrown in by an opaque-microscopic-illumination.[23] Sept. 20, 1783, Position 41° 24′ south preceding the same star; very exact, and by the same kind of illumination. ... Its diameter is about 10 or 15″. I have

[20] [M57 = NGC 6720, another large planetary (gaseous) nebula, the 'Ring Nebula in Lyra'. – D.]

[21] [In this last section Herschel correctly classifies (by the name we use today) and describes these small gaseous nebulae surrounding some hot stars. – D.]

[22] [NGC 7009. The "small star" is identifiable on modern photographs. The distance and position angle from the star are fixed, and the values show Herschel's observational error. Although he gives the distances to the nearest *third*, the scatter in his measures is almost 4,000 times greater – an example of the meaningless 'precision' of which he was sometimes guilty. – D.]

[23] It may be of use to explain this kind of illumination for which the Newtonian reflector is admirably constructed. On the side opposite the eye-piece an opening is to be made in the tube, through which the light may be thrown in, so as to fall on some reflecting body, or concave perforated mirror, within the eye-piece, that may throw it back upon the wires. By this means none of the direct rays can reach the eye, and those few which are reflected again from the wires do not interfere sensibly with the faintest objects, which may thus be seen undisturbed.

examined it with the powers of 71, 227, 278, 460, and 932; and it follows the laws of magnifying, so that its body is no illusion of light. It is a little oval, and in the 7-feet reflector pretty well defined, but not sharp on the edges. In the 20-feet, of 18.7 inch aperture, it is much better defined, and has much of a planetary appearance, being all over of an uniform brightness, in which it differs from nebulae: its light seems however to be of the starry nature, which suffers not nearly so much as the planetary disks are known to do, when much magnified.

The second of these bodies precedes the 13th of FLAMSTEED's Andromeda about 1′,6 in time, and is 22′ more south.[24] It has a round, bright, pretty well defined planetary disk of about 12″ diameter, and is a little elliptical. When it is viewed with a 7-feet reflector, or other inferior instruments, it is not nearly so well defined as with the 20-feet. Its situation with regard to a pretty considerable star is, Distance (with a compound glass of a low power) 7′ 51″ 34‴. Position 12° 0′ s. preceding. Diameter taken with 278, 14″ 42‴.

The third follows B (FL. 44) Ophiuchi 4′,1 in time, and is 23′ more north. It is round, tolerably well defined, and pretty bright; its diameter is about 30″.[25]

The fourth follows η Sagittae 17′,1 in time, and is 2′ more north. It is perfectly round, pretty bright, and pretty well defined; about ¾ min. in diameter.[26]

The fifth follows the 21st Vulpecula 2′,1 in time, and is 1° 46′ more north. It is exactly round, of an equal light throughout, but pretty faint, and about 1′ in diameter.

The sixth precedes *h* (FL. 39) Cygni 8′,1 in time, and is 1° 26′ more south. It is perfectly round, and of an equal light, but pretty faint; its diameter is near 1′, and the edges are pretty well defined.[27]

The planetary appearance of the two first is so remarkable, that we can hardly suppose them to be nebulae; their light is so uniform, as well as vivid, the diameters so small and well defined, as to make it almost improbable they should belong to that species of bodies. On the other hand, the effect of different powers seems to be much against their light's being of a planetary nature, since it preserves its brightness nearly in the same manner as the stars do in similar trials. If we would suppose them to be single stars with large diameters we shall find it difficult to account for their not being brighter; unless we should admit that the intrinsic light of some stars may be very much inferior to that of the generality, which however can hardly be imagined to extend to such a degree. We might suspect them to be comets about their aphelion, if the brightness as well as magnitude of the diameters did not oppose this idea; so that after all, we

[24] [NGC 7662 – D.] [25] [NGC 6369 – D.] [26] [NGC 6905 – D.]

[27] ["The fifth. . . . The sixth . . .". The same object twice described = NGC 6894. – D.]

can hardly find any hypothesis so probable as that of their being Nebulae; but then they must consist of stars that are compressed and accumulated in the highest degree. If it were not perhaps too hazardous to pursue a former surmise of a renewal in what I figuratively called the Laboratories of the universe, the stars forming these extraordinary nebulae, by some decay or waste of nature, being no longer fit for their former purposes, and having their projectile forces, if any such they had, retarded in each others' atmosphere, may rush at last together, and either in succession, or by one general tremendous shock, unite into a new body. Perhaps the extraordinary and sudden blaze of a new star in Cassiopeia's chair, in 1572, might possibly be of such a nature.[28] But lest I should be led too far from the path of observation, to which I am resolved to limit myself, I shall only point out a considerable use that may be made of these curious bodies. If a little attention to them should prove that, having no annual parallax, they belong most probably to the class of nebulae, they may then be expected to keep their situation better than any one of the stars belonging to our system, on account of their being probably at a very great distance. Now to have a fixed point somewhere in the heavens, to which the motions of the rest may be referred, is certainly of considerable consequence in Astronomy; and both these bodies are bright and small enough to answer that end.[29]

W. Herschel.

Datchet near Windsor,
January 1, 1785.

[28] [The "new star in Cassiopeia's chair" is the supernova of 1572. Herschel's recognition that small bright galaxies (which at this time he thought planetary nebulae to be) would be useful as providing an absolute reference frame against which to measure proper motions is far-seeing. The improvements in telescopes and practical techniques in the last sixty years have made it an important feature of present-day positional astronomy. – D]

[29] [Herschel here adds two more such objects, NGC 1535 and 3242.]

1789: "Remarks on the construction of the heavens"

Remarks on the Construction of the Heavens[1]

Read June 11, 1789.

By the continuation of a review of the heavens with my twenty-feet reflector, I am now furnished with a second thousand of new Nebulae.

These curious objects, not only on account of their number, but also in consideration of their great consequence, as being no less than whole sidereal systems, we may hope, will in future engage the attention of Astronomers. With a view to induce them to undertake the necessary observations, I offer them the following catalogue [not reproduced], which, like my former one, of which it is a continuation, contains a short description of each nebula or cluster of stars, as well as its situation with respect to some known object. . . .

The method I have taken of *analyzing* the heavens, if I may so express myself, is perhaps the only one by which we can arrive at a knowledge of their construction. In the prosecution of so extensive an undertaking, it may well be supposed that many things must have been suggested, by the great variety in the order, the size, and the compression of the stars, as they presented themselves to my view, which it will not be improper to communicate.

To begin our investigation according to some order, let us depart from the objects immediately around us to the most remote that our telescopes, of the greatest *power to penetrate into space*, can reach. We shall touch but slightly on things that have already been remarked.

From the earth, considered as a planet, and the moon as its satellite, we pass through the region of the rest of the planets, and their satellites. The similarity

[1] [The title of the complete paper is "Catalogue of a Second Thousand of new Nebulae and Clusters of Stars; with a few introductory Remarks on the Construction of the Heavens", *Phil. Trans.*, **79** (1789), 212–255.]

between all these bodies is sufficiently striking to allow us to comprehend them under one general definition, of bodies not luminous in themselves, revolving round the sun. The great diminution of light, when reflected from such bodies, especially when they are also at a great distance from the light which illuminates them, precludes all possibility of following them a great way into space. But if we did not know that light diminishes as the squares of the distances encrease, and that moreover in every reflection a very considerable part is intirely lost, the motion of comets, whereby the space through which they run is measured out to us, while on their return from the sun we see them gradually disappear as they advance towards their aphelia, would be sufficient to convince us that bodies shining only with borrowed light can never be seen at any very great distance. This consideration brings us back to the sun, as a refulgent fountain of light, whilst it establishes at the same time beyond a doubt that every star must likewise be a sun, shining by its own native brightness. Here then we come to the more capital parts of the great construction.

These suns, every one of which is probably of as much consequence to a system of planets, satellites, and comets, as our own sun, are now to be considered, in their turn, as the minute parts of a proportionally greater whole. I need not repeat that by my analysis it appears, that the heavens consist of regions where the suns are gathered into separate systems, and that the catalogues I have given comprehend a list of such systems; but may we not hope that our knowledge will not stop short at the bare enumeration of phaenomena capable of giving us so much instruction? Why should we be less inquisitive than the natural philosopher, who sometimes, even from an inconsiderable number of specimens of a plant, or an animal, is enabled to present us with the history of its rise, progress, and decay? Let us then compare together, and class some of these numerous sidereal groups, that we may trace the operations of natural causes as far as we can perceive their agency. The most simple form, in which we can view a sidereal system, is that of being globular. This also, very favourably to our design, is that which has presented itself most frequently, and of which I have given the greatest collection.

But, first of all, it will be necessary to explain what is our idea of a cluster of stars, and by what means we have obtained it. For an instance, I shall take the phaenomenon which presents itself in many clusters: It is that of a number of lucid spots, of equal lustre, scattered over a circular space, in such a manner as to appear gradually more compressed towards the middle; and which compression, in the clusters to which I allude, is generally carried so far, as, by imperceptible degrees, to end in a luminous center, of a resolvable blaze of light.[2] To

[2] [Herschel is clearly writing here of globular clusters, and goes on to show (correctly) that they are of truly spherical form. – D.]

solve this appearance, it may be conjectured, that stars of any given, very unequal magnitudes, may easily be so arranged, in scattered, much extended, irregular rows, as to produce the above described picture; or, that stars, scattered about almost promiscuously within the frustum of a given cone, may be assigned of such properly diversified magnitudes as also to form the same picture. But who, that is acquainted with the doctrine of chances, can seriously maintain such improbable conjectures? To consider this only in a very coarse way, let us suppose a cluster to consist of 5000 stars, and that each of them may be put into one of 5000 given places, and have one of 5000 assigned magnitudes. Then, without extending our calculation any further, we have five and twenty millions of chances, out of which only one will answer the above improbable conjecture, while all the rest are against it. When we now remark that this relates only to the given places within the frustum of a supposed cone, whereas these stars might have been scattered all over the visible space of the heavens; that they might have been scattered, even within the supposed cone, in a million of places different from the assumed ones, the chance of this apparent cluster's not being a real one, will be rendered so highly improbable that it ought to be intirely rejected.

Mr. MICHELL computes, with respect to the six brightest stars of the Pleiades only, that the odds are near 500000 to 1 that no six stars, out of the number of those which are equal in splendour to the faintest of them, scattered at random in the whole heavens, would be within so small a distance from each other as the Pleiades are.[3]

Taking it then for granted that the stars which appear to be gathered together in a group are in reality thus accumulated, I proceed to prove also that they are nearly of an equal magnitude. The cluster itself, on account of the small angle it subtends to the eye, we must suppose to be very far removed from us. For, were the stars which compose it at the same distance from one another as Sirius is from the sun; and supposing the cluster to be seen under an angle of 10 minutes, and to contain 50 stars in one of its diameters, we should have the mean distance of such stars twelve seconds; and therefore the distance of the cluster from us about seventeen thousand times greater than the distance of Sirius. Now, since the apparent magnitude of these stars is equal, and their distance from us is also equal, – because we may safely neglect the diameter of the cluster, which, if the center be seventeen thousand times the distance of Sirius from us, will give us seventeen thousand and twenty-five for the farthest, and seventeen thousand wanting twenty-five for the nearest star of the cluster; – it follows that we must

[3] *Phil. Trans.* vol. LVII. p. 246.

either give up the idea of a cluster, and recur to the above refuted supposition, or admit the equality of the stars that compose these clusters. It is to be remarked that we do not mean intirely to exclude all variety of size; for the very great distance, and the consequent smallness of the component clustering stars, will not permit us to be extremely precise in the estimation of their magnitudes; though we have certainly seen enough of them to know that they are contained within pretty narrow limits; and do not, perhaps, exceed each other in magnitude more than in some such proportion as one full-grown plant of a certain species may exceed another full-grown plant of the same species.

If we have drawn proper conclusions relating to the size of stars, we may with still greater safety speak of their relative situations, and affirm that in the same distances from the center an equal scattering takes place. If this were not the case, the appearance of a cluster could not be uniformly encreasing in brightness towards the middle, but would appear nebulous in those parts which were more crowded with stars; but, as far as we can distinguish, in the clusters of which we speak, every concentric circle maintains an equal degree of compression, as long as the stars are visible; and when they become too crowded to be distinguished, an equal brightness takes place, at equal distances from the center, which is the most luminous part.

The next step in my argument will be to shew that these clusters are of a globular form. This again we rest on the sound doctrine of chances. Here, by way of strength to our argument, we may be allowed to take in all round nebulae, though the reasons we have for believing that they consist of stars have not as yet been entered into. For, what I have to say concerning their spherical figure will equally hold good whether they be groups of stars or not. In my catalogues we have, I suppose, not less than one thousand of these round objects. Now, whatever may be the shape of a group of stars, or of a Nebula, which we would introduce instead of the spherical one, such as a cone, an ellipsis, a spheroid, a circle or a cylinder, it will be evident that out of a thousand situations, which the axes of such forms may have, there is but one that can answer the phaenomenon for which we want to account; and that is, when those axes are exactly in a line drawn from the object to the place of the observer. Here again we have a million of chances of which all but one are against any other hypothesis than that which we maintain, and which, for this reason, ought to be admitted.

The last thing to be inferred from the above related appearances is, that these clusters of stars are more condensed towards the center than at the surface. If there should be a group of stars in a spherical form, consisting of such as were equally scattered over all the assigned space, it would not appear to be very gradually more compressed and brighter in the middle; much less would it seem to have a bright nucleus in the center. A spherical cluster of an equal compression within, – for that such there are will be seen hereafter, – may be distinguished by the degrees of brightness which take place in going from the center to the circumference. Thus, when a is the brightness in the center, it will be $\surd(a^2 - x^2)$ at any other distance x from the center. Or, putting $a = 1$, and x = any decimal fraction; then, in a table of natural sines, where x is the sine, the brightness at x will be expressed by the cosine. Now, as a gradual encrease of brightness does not agree with the degrees calculated from a supposition of an equal scattering, and as the cluster has been proved to be spherical, it must needs be admitted that there is indeed a greater accumulation towards the center. And thus, from the above-mentioned appearances, we come to know that there are globular clusters of stars nearly equal in size, which are scattered evenly at equal distances from the middle, but with an encreasing accumulation towards the center.

We may now venture to raise a superstructure upon the arguments that have been drawn from the appearance of clusters of stars and nebulae of the form I have been examining, which is that of which I have made mention in my "*Theoretical view – Formation of Nebulae – Form I.*"[4] It is to be remarked that when I wrote the paragraph I refer to, I delineated nature as well as I do now; but, as I there gave only a general sketch, without referring to particular cases, what I then delivered may have been looked upon as little better than hypothetical reasoning, whereas in the present instance this objection is intirely removed, since actual and particular facts are brought to vouch for the truth of every inference.

Having then established that the clusters of stars of the 1st Form, and round nebulae, are of a spherical figure, I think myself plainly authorized to conclude that they are thus formed by the action of central powers. To manifest the validity of this inference, the figure of the earth may be given as an instance; whose rotundity, setting aside small deviations, the causes of which are well known, is without hesitation allowed to be a phaenomenon decisively establishing a centipetal force. Nor do we stand in need of the revolving satellites of Jupiter, Saturn, and the Georgium Sidus [Uranus], to assure us that the same powers are likewise lodged in the masses of these planets. Their globular figure alone must be admitted as a sufficient argument to render this point uncontrovertible. We also apply this inference with equal propriety to the body of the sun, as well as to that of Mercury, Venus, Mars, and the Moon; as owing their spherical shape to the same cause. And how can we avoid inferring, that the

[4] *Phil. Trans.* vol. LXXV. p. 214 [above, p. 115].

construction of the clusters of stars, and nebulae likewise, of which we have been speaking, is as evidently owing to central powers?

Besides, the step that I here make in my inference is in fact a very easy one, and such as ought freely to be granted. Have I not already shewn that these clusters cannot have come to their present formation by any random scattering of stars? The doctrine of chance, by exposing the very great odds against such hypotheses, may be said to demonstrate that the stars are thus assembled by some power or other. Then, what do I attempt more than merely to lead the mind to the conditions under which this power is seen to act?

In a case of such consequence I may be permitted to be a little more diffuse, and draw additional arguments from the internal construction of spherical clusters and nebulae. If we find that there is not only a general form, which, as has been proved, is a sufficient manifestation of a centripetal force, what shall we say when the accumulated condensation, which every where follows a direction towards a center, is even visible to the very eye? Were we not already acquainted with attraction, this gradual condensation would point out a central power, by the remarkable disposition of the stars tending towards a center. In consequence of this visible accumulation, whether it may be owing to attraction only, or whether other powers may assist in the formation, we ought not to hesitate to ascribe the effect to such as are *central*; no phaenomena being more decisive in that particular, than those of which I am treating.

I am fully aware of the consequences I shall draw upon myself in but mentioning other powers that might contribute to the formation of clusters. A mere hint of this kind, it will be expected, ought not to be given without sufficient foundation; but let it suffice at present to remark that my arguments cannot be affected by my terms: whether I am right to use the plural number, – central powers, – or whether I ought only to say, – the known central force of gravity, – my conclusions will be equally valid. I will however add, that the idea of other central powers being concerned in the construction of the sidereal heavens, is not one that had only lately occurred to me. Long ago I have entertained a certain theory of diversified central powers of attractions and repulsions; an exposition of which I have even delivered in the years 1780, and 1781, to the Philosophical Society then existing at Bath, in several mathematical papers upon that subject. I shall, however, set aside an explanation of this theory, which would not only exceed the intended limits of this paper, but is moreover not required for what remains at present to be added, and therefore may be given some other time, when I can enter more fully into the subject of the interior construction of sidereal systems.

To return, then, to the case immediately under our present consideration, it will be sufficient that I have abundantly proved that the formation of round clusters of stars and nebulae is either owing to central powers, or at least to one such force as refers to a center.

I shall now extend the weight of my argument, by taking in likewise every cluster of stars or nebula that shews a gradual condensation, or encreasing brightness, towards a center or certain point; whether the outward shape of such clusters or nebulae be round, extended, or of any other given form. What has been said with regard to the doctrine of chance, will of course apply to every cluster, and more especially to the extended and irregular shaped ones, on account of their greater size: It is among these that we find the largest assemblages of stars, and most diffusive nebulosities; and therefore the odds against such assemblages happening without some particular power to gather them, encrease exceedingly with the number of the stars that are taken together. But if the gradual accumulation either of stars or encreasing brightness has before been admitted as a direction to the seat of power, the same effect will equally point out the same cause in the cases now under consideration. There are besides some additional circumstances in the appearance of extended clusters and nebulae, that very much favour the idea of a power lodged in the brightest part. Although the form of them be not globular, it is plainly to be seen that there is a tendence towards sphericity, by the swell of the dimensions the nearer we draw towards the most luminous place, denoting as it were a course, or tide of stars, setting towards a center. And – if allegoral expressions may be allowed – it should seem as if the stars thus flocking towards the seat of power were stemmed by the crowd of those already assembled, and that while some of them are successful in forcing their predecessors sideways out of their places, others are themselves obliged to take up with lateral situations, while all of them seem equally to strive for a place in the central swelling, and generating spherical figure.

Since then almost all the nebulae and clusters of stars I have seen, the number of which is not less than three and twenty hundred, are more condensed and brighter in the middle; and since, from every form, it is now equally apparent that the central accumulation or brightness must be the result of central powers, we may venture to affirm that this theory is no longer an unfounded hypothesis, but is fully established on grounds which cannot be overturned.

Let us endeavour to make some use of this important view of the constructing cause, which can thus model sidereal systems. Perhaps, by placing before us the very extensive and varied collection of clusters, and nebulae furnished by my catalogues, we may be able to trace the progress of its operation, in the great laboratory of the Universe.

If these clusters and nebula were all of the same shape, and had the same gradual condensation, we should make but little progress in this inquiry; but, as we find so great a variety in their appearances, we shall be much sooner at a loss how to account for such various phaenomena, than be in want of materials upon which to exercise our inquisitive endeavours.

Some of these round clusters consist of stars of a certain magnitude, and given degree of compression, while the whole cluster itself takes up a space of perhaps 10 minutes; others appear to be made up of stars that are much smaller, and much more compressed, when at the same time the cluster itself subtends a much smaller angle, such as 5 minutes. This diminution of the apparent size, and compression of stars, as well as diameter of the cluster to 4, 3, 2 minutes, may very consistently be ascribed to the different distances of these clusters from the place in which we observe them; in all which cases we may admit a general equality of the sizes, and compression of the stars that compose them, to take place. It is also highly probable that a continuation of such decreasing magnitudes, and encreasing compression, will justly account for the appearance of round, easily resolvable, nebulae; where there is almost a certainty of their being clusters of stars. And no Astronomer can hesitate to go still farther, and extend his surmises by imperceptible steps to other nebulae, that still preserve the same characteristics, with the only variations of vanishing brightness, and reduction of size.

Other clusters there are that, when they come to be compared with some of the former, seem to contain stars of an equal magnitude, while their compression appears to be considerably different. Here the supposition of their being at different distances will either not explain the apparently greater compression, or, if admitted to do this, will convey to us a very instructive consequence: which is, that the stars which are thus supposed not to be more compressed than those in the former cluster, but only to appear so on account of their greater distance, must needs be proportionally larger, since they do not appear of less magnitude than the former. As therefore, one or other of these hypotheses must be true, it is not [at] all improbable but that, in some instances, the stars may be more compressed; and in others, of a greater magnitude. This variety of size, in different spherical clusters, I am however inclined to believe, may not go farther than the difference in size, found among the individuals belonging to the same species of plants, or animals, in their different states of age, or vegetation, after they are come to a certain degree of growth. A farther inquiry into the circumstance of the extent, both of condensation and variety of size, that may take place with the stars of different clusters, we shall postpone till other things have been previously discussed.

Let us then continue to turn our view to the power which is moulding the different assortments of stars into spherical clusters. Any force, that acts uninterruptedly, must produce effects proportional to the time of its action. Now, as it has been shewn that the spherical figure of a cluster of stars is owing to central powers, it follows that those clusters which, *ceteris paribus*, are the most compleat in this figure, must have been the longest exposed to the action of these causes.[5] This will admit of various points of view. Suppose for instance that 5000 stars had been once in a certain scattered situation, and that other 5000 equal stars had been in the same situation, then that of the two clusters which had been longest exposed to the action of the modelling power, we suppose, would be most condensed, and more advanced to the maturity of its figure. An obvious consequence that may be drawn from this consideration is, that we are enabled to judge of the relative age, maturity, or climax of a sidereal system, from the disposition of its component parts; and, making the degrees of brightness in nebulae stand for the different accumulation of stars in clusters, the same conclusions will extend equally to them all. But we are not to conclude from what has been said that every spherical cluster is of an equal standing in regard to absolute duration, since one that is composed of a thousand stars only, must certainly arrive to the perfection of its form sooner than another, which takes in a range of a million. Youth and age are comparative expressions; and an oak of a certain age may be called very young, while a contemporary shrub is already on the verge of its decay. The method of judging with some assurance of the condition of any sidereal system may perhaps not improperly be drawn from the standard laid down; so that, for instance, a cluster or nebula which is very gradually more compressed and bright towards the middle, may be in the perfection of its growth, when another which approaches to the condition pointed out by a more equal compression, such as the nebulae I have called *Planetary* seem to present us with, may be looked upon as very aged, and drawing on towards a period of change, or dissolution. This has been before surmised, when, in a former paper, I considered the uncommon degree of compression that must prevail in a nebula to give it a planetary aspect; but the argument, which is now drawn from the powers that have collected the formerly scattered stars to the form we find they have assumed, must greatly corroborate that sentiment.

[5] [Watson made the sensible comment that Herschel's work on the age of clusters of stars "requires *stronger* premises to warrant the conclusions, especially as you do not set out with the probable primeval arrangement of Stars, but it may still serve to shew the thing hypothetically, or you might shew the Chances to be on that side. For tho' I may not be able to produce any reason why irregular clusters have not become round; it should seem you ought to shew that the previous arrangement in such clusters was similar to that of the globular ones & that therefore time only was required for the effect of making them globular likewise" (letter of 12 May 1789, RAS W.1/13.W.55).]

This method of viewing the heavens seems to throw them into a new kind of light. They now are seen to resemble a luxuriant garden, which contains the greatest variety of productions, in different flourishing beds; and one advantage we may at least reap from it is, that we can, as it were, extend the range of our experience to an immense duration. For, to continue the simile I have borrowed from the vegetable kingdom, is it not almost the same thing, whether we live successively to witness the germination, blooming, foliage, fecundity, fading, withering, and corruption of a plant, or whether a vast number of specimens, selected from every stage through which the plant passes in the course of its existence, be brought at once to our view?

WILLIAM HERSCHEL.

Slough near Windsor, May 1, 1789.

1791: “On nebulous stars”

On Nebulous Stars, properly so called[1]

Read February 10, 1791.

In one of my late examinations of a space in the heavens, which I had not reviewed before, I discovered *a star of about the 8th magnitude, surrounded with a faintly luminous atmosphere, of a considerable extent.*[2] The phaenomenon was so striking that I could not help reflecting upon the circumstances that attended it, which appeared to me to be of a very instructive nature, and such as may lead to inferences which will throw a considerable light on some points relating to the construction of the heavens.

Cloudy or nebulous stars have been mentioned by several astronomers; but this name ought not to be applied to the objects which they have pointed out as such; for, on examination, they proved to be either mere clusters of stars, plainly to be distinguished with my large instruments, or such nebulous appearances as might be reasonably supposed to be occasioned by a multitude of stars at a vast distance. The milky way itself, as I have shewn in some former Papers, consists intirely of stars, and by imperceptible degrees I have been led on from the most evident congeries of stars to other groups in which the lucid points were smaller, but still very plainly to be seen; and from them to such wherein they could but barely be suspected, till I arrived at last to spots in which no trace of a star was to be discerned. But then the gradations to these latter were by such well-connected steps as left no room for doubt but that all these phaenomena were equally occasioned by stars, variously dispersed in the immense expanse of the universe.

1 [*Phil. Trans.*, **81** (1791), 71–88.]

2 [This object Herschel identifies as his IV. 69 = NGC 1514, a planetary nebula (but for him “a nebulous star”, see p. 153). – D.]

When I pursued these researches, I was in the situation of a natural philosopher who follows the various species of animals and insects from the height of their perfection down to the lowest ebb of life; when, arriving at the vegetable kingdom, he can scarcely point out to us the precise boundary where the animal ceases and the plant begins; and may even go so far as to suspect them not to be essentially different. But recollecting himself, he compares, for instance, one of the human species to a tree, and all doubt upon the subject vanishes before him. In the same manner we pass through gentle steps from a coarse cluster of stars, such as the Pleiades, the Praesepe, the milky way, the cluster in the Crab,[3] the nebula in Hercules,[4] that near the preceding hip of Bootes,[5] the 17th, 38th, 41st of the 7th class of my Catalogues, the 10th, 20th, 35th of the 6th class, the 33d, 48th, 213th of the 1st, the 12th, 150th, 756th of the 2d, and the 18th, 140th, 725th of the 3d, without any hesitation, till we find ourselves brought to an object such as the nebula in Orion, where we are still inclined to remain in the once adopted idea, of stars exceedingly remote, and inconceivably crowded, as being the occasion of that remarkable appearance.[6] It seems, therefore, to

[3] ["the cluster in the Crab": probably M67 = NGC 2682, to which Herschel has referred earlier (p. 111) as part of the "stratum of Cancer". – D.]

[4] ["nebula in Hercules", M13 = NGC 6205, a globular cluster. – D.]

[5] ["that near the preceding hip of Bootes", M3 = NGC 5272, a globular cluster. – D.]

[6] ["the 7th class of my Catalogues". These classes are: (I) Bright nebulae; (II) Faint nebulae; (III) Very faint nebulae; (IV) Planetary nebulae, &c.; (V) Very large nebulae; (VI) Very condensed and rich clusters of stars; (VII) Pretty much compressed clusters of large or small stars; (VIII) Coarsely scattered clusters of stars.

H.VII.	17	= NGC 2362	Galactic cluster
	38	2324	Galactic cluster
	41	7296	Galactic cluster
H.VI.	10	6144	Globular cluster
	20	288	Globular cluster
	35	136	Small compact galactic cluster
H. I.	33	4124	Galaxy
	48	6356	Globular cluster
	213	4449	Bright elongated galaxy, Type Irr 1
H. II.	12	4651	Galaxy
	150	5645	Galaxy, Irr
	756	5820	Galaxy, S0
H.III.	18	4505	Galaxy
	140	6064	Galaxy
	725	4242	Galaxy

Note here the order in which Herschel arranges his objects, from open clusters like the Pleiades through classes VII, VI, I, II, to III, ending with the Orion Nebula which he still believes to be made of stars "exceedingly remote, and inconceivably crowded". (In footnotes, not reproduced, Herschel gives the positions of these nebulae.) – D.]

require a more dissimilar object to set us right again. A glance like that of the naturalist, who casts his eye from the perfect animal to the perfect vegetable, is wanting to remove the veil from the mind of the astronomer. The object I have mentioned above, is the phaenomenon that was wanting for this purpose. View, for instance, the 19th cluster of my 6th class,[7] and afterwards cast your eye on this cloudy star,[8] and the result will be no less decisive than that of the naturalist we have alluded to. Our judgement, I may venture to say, will be, that *the nebulosity about the star is not of a starry nature*.

But, that we may not be too precipitate in these new decisions, let us enter more at large into the various grounds which induced us formerly to surmise, that every visible object, in the extended and distant heavens, was of the starry kind, and collate them with those which now offer themselves for the contrary opinion.

It has been observed, on a former occasion, that all the smaller parts of other great systems, such as the planets, their rings and satellites, the comets, and such other bodies of the like nature as may belong to them, can never be perceived by us, on account of the faintness of light reflected from small, opaque objects; in my present remarks, therefore, all these are to be intirely set aside.

A well connected series of objects, such as we have mentioned above, has led us to infer, that all nebulae consist of stars. This being admitted, we were authorized to extend our analogical way of reasoning a little farther. Many of the nebulae had no other appearance than that whitish cloudiness, on the blue ground upon which they seemed to be projected; and why the same cause should not be assigned to explain the most extensive nebulosities, as well as those that amounted only to a few minutes of a degree in size, did not appear. It could not be inconsistent to call up a telescopic milky way, at an immense distance, to account for such phaenomena; and if any part of the nebulosity seemed detached from the rest, or contained a visible star or two, the probability of seeing a few near stars, apparently scattered over the far distant regions of myriads of sidereal collections, rendered nebulous by their distance, would also clear up these singularities.

In order to be more easily understood in my remarks on the comparative disposition of the heavenly bodies, I shall mention some of the particulars which introduced the ideas of *connection* and *disjunction*: for these, being properly founded

[7] [H. VI. 19 = NGC 5897, a globular cluster. – D.]

[8] ["this cloudy star". The key object of this paper. Herschel gives its position later in the paper; its identification is H. IV. 69 = NGC 1514. It is a planetary nebula, measuring about $2' \times 1'.5$, with a fairly conspicuous central star of $9^{m}.7$. The surface brightness of the nebulosity is smaller, and the central star relatively more prominent, than is usual in planetary nebulae. – D.]

upon an examination of objects that may be reviewed at any time, will be of considerable importance to the validity of what we may advance with regard to my lately discovered nebulous stars.

On June the 27th, 1786, I saw a beautiful cluster of very small stars of various sizes, about 15′ in diameter, and very rich of stars.[9] On viewing this object, it is impossible to withhold our assent to the idea which occurs, that these stars are connected so far one with another as to be gathered together, within a certain space, of little extent, when compared to the vast expanse of the heavens. As this phaenomenon has been repeatedly seen in a thousand cases, I may justly lay great stress on the idea of such stars being connected.

In the year 1779, the 9th of September, I discovered a very small star near ε Bootis.[10] The question here occurring, whether it had any connection with ε or not, was determined in the negative; for, considering the number of stars scattered in a variety of places, it is very far from being uncommon, that a star at a great distance should happen to be nearly in a line drawn from the sun through ε, and thus constitute the observed double star.

The 7th of September, 1782, when I first saw the planetary nebula near ν Aquarii,[11] I pronounced it to be a system whose parts were connected together. Without entering into any kind of calculation, it is evident, that a certain equal degree of light within a very small space, joined to the particular shape this object presents to us, which is nearly round, and even in its deviation consistent with regularity, being a little elliptical, ought naturally to give us the idea of a conjunction in the things that produce it. And a considerable addition to this argument may be derived from a repetition of the same phaenomenon, in nine or ten more of a similar construction.

When I examined the cluster of stars, following the head of the great dog,[12] I found on the 19th of March, 1786, that there was within this cluster a round, resolvable nebula, of about two minutes in diameter, and nearly of an equal degree of light throughout.[13] Here, considering that the cluster was free from nebulosity in other parts, and that many such clusters, as well as many such nebulae, exist in divers parts of the heavens, it appeared to me very probable, that the nebula was unconnected with the cluster; and that a similar reason would as easily account for this appearance as it had resolved the phaenomenon of the double star near ε Bootis; that is, a casual situation of our sun and the two

[9] [H. VI. 23 = NGC 6645, a galactic cluster. – D.]

[10] Phil. Trans. Vol. LXXII. p. 115. Catalogue of Double Stars, I. 1.

[11] [H. IV.1 = NGC 7009, planetary nebula correctly described by Herschel. – D.]

[12] [M46 = NGC 2437, galactic cluster. – D.]

[13] [H. IV. 39 = NGC 2438, planetary nebula involved in M46. – D.]

other objects nearly in a line. And though it may be rather more remarkable, that this should happen with two compound systems, which are not by far so numerous as single stars, we have, to make up for this singularity, a much larger space in which it may take place, the cluster being of a very considerable extent.

On the 15th of February, 1786, I discovered that one of my planetary nebulae,[14] had a spot in the center, which was more luminous than the rest, and with long attention, a very bright, round, well defined center became visible. I remained not a single moment in doubt, but that the bright center was connected with the rest of the apparent disk.

In the year 1785, the 6th of October, I found a very bright, round nebula, of about 1 minute in diameter.[15] It has a large, bright nucleus in the middle, which is undoubtedly connected with the luminous parts about it. And though we must confess, that if this phaenomenon, and many more of the same nature, recorded in my catalogues of nebulae, consist of clustering stars, we find ourselves involved in some difficulty to account for the extraordinary condensation of them about the center; yet the idea of a connection between the outward parts and these very condensed ones within is by no means lessened on that account.

There is a telescopic milky way, which I have traced out in the heavens in many sweeps made from the year 1783 to 1789.[16] It takes up a space of more than 60 square degrees of the heavens, and there are thousands of stars scattered over it: among others, four that form a trapezium, and are situated in the well known nebula of Orion, which is included in the above extent. All these stars, as well as the four I have mentioned, I take to be intirely unconnected with the nebulosity which involves them in appearance. Among them is also *d* Orionis, a cloudy star, improperly so called[17] by former astronomers; but it does not seem to be connected with the milkiness any more than the rest.

[14] [H. IV. 37 = NGC 6543, planetary nebula. – D.]

[15] [H. I. 107 = NGC 1407, a galaxy of type E0 with a pronounced bright nucleus – a not unusual feature of galaxies. – D.]

[16] R.A. from 5 h. 15′ 8″ to 5 h. 39′ 1″. P.D. from 87° 46′ to 98° 10′. [The area described by H. is extended N–S and includes the Orion Nebula and the three stars of Orion's Belt. The area is in the Milky Way, rich in nebulosity and highly luminous stars of types O and B, but there are no features defining its boundary that would justify Herschel's very accurate specification of its limits in R.A. and Dec. We now know that for the greater part the stars and gas in such regions are physically associated. – D.]

[17] [*d* Orionis, "a cloudy star, improperly so called", = 49 Ori = BD–7°1142. There are very faint gaseous nebulae in the vicinity, including IC 430, but if Herschel did not see them, certainly his predecessors could not have done. – D.]

I come now to some other phaenomena, that, from their singularity, merit undoubtedly a very full discussion. Among the reasons which induced us to embrace the opinion, that all very faint milky nebulosity ought to be ascribed to an assemblage of stars is, that we could not easily assign any other cause of sufficient importance for such luminous appearances, to reach us at the immense distance we must suppose ourselves to be from them. But if an argument of considerable force should now be brought forward, to shew the existence of a luminous matter, in a state of modification very different from the construction of a sun or star, all objections, drawn from our incapacity of accounting for new phaenomena upon old principles, will lose their validity.

Hitherto I have been shewing, by various instances in objects whose places are given, in what manner we may form the ideas of connection and its contrary by an attentive inspection of them only: I will now relate a series of observations, with remarks upon them as they are delivered, from which I shall afterwards draw a few simple conclusions, that seem to be of considerable importance.

To distinguish the observations from the remarks, the former are given in italics, and the date annexed is that on which the objects were discovered; but the descriptions are extracted from all the observations that have been made upon them.

October 16, 1784. *A star of about the 9th magnitude, surrounded by a milky nebulosity, or chevelure, of about 3 minutes in diameter. The nebulosity is very faint, and a little extended or elliptical, the extent being not far from the meridian, or a little from north preceding to south following. The chevelure involves a small star, which is about 1½ minute north of the cloudy star; other stars of equal magnitude are perfectly free from this appearance.*[18]

My present judgement concerning this remarkable object is, that the nebulosity belongs to the star which is situated in its center. The small one, on the contrary, which is mentioned as involved, being one of many that are profusely scattered over this rich neighbourhood, I suppose to be quite unconnected with this phaenomenon. A circle of three minutes in diameter is sufficiently large to admit another small star, without any bias to the judgement I form concerning the one in question.

It must appear singular, that such an object should not have immediately suggested all the remarks contained in this Paper; but about things that appear new we ought not to form opinions too hastily, and my observations on the construction of the heavens were then but entered upon. In this case, therefore, it was the safest way to lay down a rule not to reason upon the phaenomena that

[18] [H. IV. 19 = NGC 2170. Not a planetary nebula, but a star embedded in irregular galactic nebulosity, probably partly reflection nebulosity. – D.]

might offer themselves, till I should be in possession of a sufficient stock of materials to guide my researches. . . .[19]

January the 6th, 1785. *A bright star with a considerable milky chevelure; a little extended, 4 or 5′ in length, and near 4′ broad; it loses itself insensibly. Other stars of equal magnitude are perfectly free from this chevelure.*[20]

The connection between the star and the chevelure cannot be doubted, from the insensible gradation of its luminous appearance, decreasing as it receded from the center. . . .[21]

November 13, 1790. *A most singular phaenomenon! A star of about the 8th magnitude, with a faint luminous atmosphere, of a circular form, and of about 3′ in diameter. The star is perfectly in the center, and the atmosphere is so diluted, faint, and equal throughout, that there can be no surmise of its consisting of stars; nor can there be a doubt of the evident connection between the atmosphere and the star. Another star not much less in brightness, and in the same field with the above, was perfectly free from any such appearance.*[22]

This last object is so decisive in every particular, that we need not hesitate to admit it as a pattern, from which we are authorized to draw the following important consequences.

Supposing the connection between the star and its surrounding nebulosity to be allowed, we argue, that one of the two following cases must necessarily be admitted. In the first place, if the nebulosity consist of stars that are very remote, which appear nebulous on account of the small angles their mutual distances subtend at the eye, whereby they will not only, as it were, run into one another, but also appear extremely faint and diluted; then, what must be the enormous size of the central point, which outshines all the rest in so superlative a degree as to admit of no comparison? In the next place, if the star be no bigger than common, how very small and compressed must be those other luminous points that are the occasion of the nebulosity which surrounds the central one? As, by the former supposition, the luminous central point must far exceed the standard of what we call a star, so, in the latter, the shining matter about the center

19 [Omitted here are two objects of which Herschel admits "the connection between the stars and the nebulosity is not so evident as to amount to conviction".]

20 [H. IV. 24 = NGC 2023. A star embedded in a small bright patch of gaseous nebulosity, in an area following the well-known 'Horse Head' dark nebula S of ζ Orionis. In this section of his paper Herschel describes some dozen objects as examples of nebulous stars he has seen in his sweeps: this one only is given here, as an example. For the greater part he chooses correctly either planetary nebulae, or stars involved in small patches of galactic nebulosity. Many of the latter are in Orion and Monoceros, regions containing many such objects. In a few cases Herschel errs and describes a galaxy with a bright nucleus or a foreground star. – D.]

21 [Omitted here are some thirteen further examples of stars linked to nebulosity.]

22 [NGC 1514. – D.]

will be much too small to come under the same denomination; we therefore either have a central body which is not a star, or have a star which is involved in a shining fluid, of a nature totally unknown to us.

I can adopt no other sentiment than the latter, since the probability is certainly not for the existence of so enormous a body as would be required to shine like a star of the 8th magnitude, at a distance sufficiently great to cause a vast system of stars to put on the appearance of a very diluted, milky nebulosity.

But what a field of novelty is here opened to our conceptions! A shining fluid, of a brightness sufficient to reach us from the remote regions of a star of the 8th, 9th, 10th, 11th, or 12th magnitude, and of an extent so considerable as to take up 3, 4, 5, or 6 minutes in diameter! Can we compare it to the coruscations of the electrical fluid in the aurora borealis? Or to the more magnificent cone of the zodiacal light as we see it in spring or autumn? The latter, notwithstanding I have observed it to reach at least 90 degrees from the sun, is yet of so little extent and brightness as probably not to be perceived even by the inhabitants of Saturn or the Georgian planet [Uranus], and must be utterly invisible at the remoteness of the nearest fixed star.

More extensive views may be derived from this proof of the existence of a shining matter. Perhaps it has been too hastily surmised that all milky nebulosity, of which there is so much in the heavens, is owing to starlight only. These nebulous stars may serve as a clue to unravel other mysterious phaenomena. If the shining fluid that surrounds them is not so essentially connected with these nebulous stars but that it can also exist without them, which seems to be sufficiently probable, and will be examined hereafter, we may with great facility explain that very extensive, telescopic nebulosity, which, as I mentioned before, is expanded over more than sixty degrees of the heavens, about the constellation of Orion; a luminous matter accounting much better for it than clustering stars at a distance. In this case we may also pretty nearly guess at its situation, which must commence somewhere about the range of the stars of the 7th magnitude, or a little farther from us, and extend unequally in some places perhaps to the regions of those of the 9th, 10th, 11th, and 12th. The foundation for this surmise is, that, not unlikely, some of the stars that happen to be situated in a more condensed part of it, or that perhaps by their own attraction draw together some quantity of this fluid greater than what they are intitled to by their situation in it, will, of course, assume the appearance of cloudy stars; and many of those I have named are either in this stratum of luminous matter, or very near it.

We have said above, that in nebulous stars the existence of the shining fluid does not seem to be so essentially connected with the central points that it might not also exist without them. For this opinion we may assign several reasons. One of them is the great resemblance between the chevelure of these stars and the diffused extensive nebulosity mentioned before, which renders it highly probable that they are of the same nature. Now, if this be admitted, the separate existence of the luminous matter, or its independence on a central star, is fully proved. We may also judge, very confidently, that the light of this shining fluid is no kind of reflection from the star in the center; for, as we have already observed, reflected light could never reach us at the great distance we are from such objects.[23] Besides, how impenetrable would be an atmosphere of a sufficient density to reflect so great a quantity of light? And yet we observe, that the outward parts of the chevelure are nearly as bright as those that are close to the star; so that this supposed atmosphere ought to give no obstruction to the passage of the central rays. If, therefore, this matter is self-luminous, it seems more fit to produce a star by its condensation than to depend on the star for its existence.

Many other diffused nebulosities, besides that about the constellation of Orion, have been observed or suspected; but some of them are probably very distant, and run out far into space. For instance, about 5 minutes in time preceding ξ Cygni, I suspect as much of it as covers near four square degrees;[24] and much about the same quantity 44′ preceding the 125 Tauri.[25] A space of almost 8 square degrees, 6′ preceding α Trianguli, seems to be tinged with milky nebulosity.[26] Three minutes preceding the 46 Eridani, strong, milky nebulosity is expanded over more than two square degrees.[27] 54′ preceding the 13th Canum venaticorum, and again 48′ preceding the same star,[28] I found the field of view affected with whitish nebulosity throughout the whole breadth of the sweep, which was 2° 39′. 4′ following the 57 Cygni, a considerable space is filled with faint, milky nebulosity, which is pretty bright in some places, and contains the 37th nebula of my Vth class, in the brightest part of it.[29] In the neighbourhood of the 44th

[23] [Herschel has correctly asserted in the preceding paragraph but one that the zodiacal light would be invisible from the nearest star. But here he is incorrect: *reflection nebulae* are found around highly luminous stars, their light being starlight reflected from interstellar particles. Even in the commoner gaseous nebulae the gas is not of course truly self-luminous, but derives its energy from the ultraviolet light of nearby stars, or from collisional processes. – D.]

[24] [5^{m}p. ξ Cygni. The North America nebula = NGC 7000, see below. – D.]

[25] [44^{m}p. 125 Tauri. The region of the galactic nebulosity IC 2087 is 60^{m}p; Herschel's area, following an obscured region, does not show any nebulosity. – D.]

[26] [6^{m}p. α Trianguli. M33 (bright, large galaxy) is 18^{m}p, no other nebulosity. – D.]

[27] [3^{m}p. 46 Eridani. The excessively faint nebulosity in this area could not have been seen visually. – D.]

[28] [54^{m}p. 13 Canum Venaticorum. This star is now designated 31 Com. Ber. The field is devoid of nebulosity. – D.]

[29] [4^{m}f. 57 Cygni. The same field described a few lines earlier as 5^{m}p. ξ Cygni. This is the so-called "North America" nebula (diffuse gaseous), from its shape. Notoriously difficult to see visually, though photographed easily, Herschel's description of it as "pretty bright

Piscium, very faint nebulosity appears to be diffused over more than 9 square degrees of the heavens.[30] Now, all these phaenomena, as we have already seen, will admit of a much easier explanation by a luminous fluid than by stars at an immense distance.[31]

The nature of planetary nebulae, which has hitherto been involved in much darkness, may now be explained with some degree of satisfaction, since the uniform and very considerable brightness of their apparent disk accords remarkably well with a much condensed, luminous fluid; whereas to suppose them to consist of clustering stars will not so completely account for the milkiness or soft tint of their light, to produce which it would be required that the condensation of the stars should be carried to an almost inconceivable degree of accumulation. The surmise of the regeneration of stars, by means of planetary nebulae, expressed in a former Paper, will become more probable, as all the luminous matter contained in one of them, when gathered together into a body of the size of a star, would have nearly such a quantity of light as we find the planetary nebulae to give. To prove this experimentally, we may view them with a telescope that does not magnify sufficiently to shew their extent, by which means we shall gather all their light together into a point, when they will be found to assume the appearance of small stars; that is, of stars at the distance of those which we call of the 8th, 9th, or 10th magnitude. Indeed this idea is greatly supported by the discovery of a well defined, lucid point, resembling a star, in the center of one of them: for the argument which has been used, in the case of nebulous stars, to shew the probability of the existence of a luminous matter, which rested upon the disparity between a bright point and its surrounding shining fluid, may here be alleged with equal justice. If the point be a generating star, the further accumulation of the already much condensed, luminous matter, may complete it in time.

How far the light that is perpetually emitted from millions of suns may be concerned in this shining fluid, it might be presumptuous to attempt to determine; but, notwithstanding the unconceivable subtilty of the particles of light, when the number of the emitting bodies is almost infinitely great, and the time of the continual emission indefinitely long, the quantity of emitted particles

in some places", as seen with a speculum metal mirror of long focus, testifies to his visual acuity. – D.]

30 [Neighbourhood of 44 Piscium. No nebulosity exists here: the field is in the ecliptic and Herschel may conceivably have seen the zodiacal band. – D.]

31 [For the greater part it is difficult to understand these six examples of "fields affected with whitish nebulosity". Except for that of the North America nebula the fields are not distinguished by any remarkable nebulosity, or are actually quite free from any. There are much better examples of the phenomenon in areas of the Milky Way, with which Herschel must have been familiar. – D.]

may well become adequate to the constitution of a shining fluid, or luminous matter, provided a cause can be found that may retain them from flying off, or reunite them. But such a cause cannot be difficult to guess at, when we know that light is so easily reflected, refracted, inflected, and deflected; and that, in the immense range of its course, it must pass through innumerable systems, where it cannot but frequently meet with many obstacles to its rectilinear progression. Not to mention the great counteraction of the united attractive force of whole sidereal systems, which must be continually exerting their power upon the particles while they are endeavouring to fly off. However, we shall lay no stress upon a surmise of this kind, as the means of verifying it are wanting: nor is it of any immediate consequence to us to know the origin of the luminous matter. Let it suffice, that its existence is rendered evident, by means of nebulous stars.

I hope it will be found, that in what has been said I have not launched out into hypothetical reasonings; and that facts have all along been kept sufficiently in view. But, in order to give every one a fair opportunity to follow me in the reflections I have been led into, the place of every object from which I have argued has been purposely added, that the validity of what I have advanced might be put to the proof by those who are inclined, and furnished with the necessary instruments to undertake an attentive and repeated inspection of the same phaenomena.

W. HERSCHEL.

Slough, Jan. 1, 1791.

1802: "Remarks on the construction of the heavens"

Remarks on the Construction of the Heavens[1]

Read July 1, 1802.

SINCE the publication of my former two catalogues of nebulae, I have, in the continuation of my telescopic sweeps, met with a number of objects that will enrich our natural history, as it may be called, of the heavens. A catalogue of them [not reproduced] will be found at the end of this paper, containing 500 new nebulae, nebulous stars, planetary nebulae, and clusters of stars. These objects have been arranged in eight classes, in conformity with the former catalogues, of which the present one is therefore a regular continuation. This renders it unnecessary to give any further explanation, either of the contents of its columns, or the abbreviations which have been used in the description of the objects.

It has hitherto been the chief employment of the physical astronomer, to search for new celestial objects, whatsoever might be their nature or condition; but our stock of materials is now so increased, that we should begin to arrange them more scientifically. The classification adopted in my catalogues, is little more than an arrangement of the objects for the convenience of the observer, and may be compared to the disposition of the books in a library, where the different sizes of the volumes is often more considered than their contents. But here, in dividing the different parts of which the sidereal heavens are composed into proper classes, I shall have to examine the nature of the various celestial objects that have been hitherto discovered, in order to arrange them in a manner most conformable to their construction. This will bring on some

[1] [The complete title of the paper is "Catalogue of 500 New Nebulae, Nebulous Stars, Planetary Nebulae, and Clusters of Stars; with Remarks on the Construction of the Heavens", *Phil. Trans.*, **92** (1802), 477–528.]

extensive considerations, which would be too long for the compass of a single paper; I shall therefore now only give an enumeration of the species that offer themselves already to our view, and leave a particular examination of the separate divisions, for some early future occasions.

In proceeding from the most simple to the more complex arrangements, several methods, taken from the known laws of gravitation, will be suggested, by which the various systems under consideration may be maintained; but here also we shall confine ourselves to a general review of the subject, as observation must furnish us first with the necessary data, to establish the application of any one of these methods on a proper foundation.

Enumeration of the parts that enter into the construction of the heavens

I. *Of insulated Stars*

In beginning our proposed enumeration, it might be expected that the solar system would stand foremost in the list; whereas, by treating of insulated stars, we seem, as it were, to overlook one of the great component parts of the universe. It will, however, soon appear that this very system, magnificent as it is, can only rank as a single individual belonging to the species which we are going to consider.

By calling a star insulated, I do not mean to denote its being totally unconnected with all other stars or systems; for no one, by the laws of gravitation, can be intirely free from the influence of other celestial bodies. But, when stars are situated at such immense distances from each other as our sun, Arcturus, Capella, Lyra [Vega], Sirius, Canopus, Markab, Bellatrix, Menkar, Shedir, Algorah, Propus, and numberless others probably are, we may then look upon them as sufficiently out of the reach, of mutual attractions, to deserve the name of insulated stars.

In order not to take this assertion for granted, without some examination, let us admit, as is highly probable, that the whole orbit of the earth's annual motion does not subtend more than an angle of one second of a degree,[2] when seen from Sirius. In consequence of this, it appears by computation, that our sun and Sirius, if we suppose their masses to be equal, would not fall together in less than 33 millions of years, even though they were not impeded by many contrary attractions of other neighbouring insulated stars; and that, consequently, with the assistance of the opposite energies exerted by such surrounding stars, these two bodies may remain for millions of ages, in a state almost equal to

[2] [The value is in fact 0.76 secs.]

undisturbed rest. A star thus situated may certainly deserve to be called insulated, since it does not immediately enter into connection with any neighbouring star; and it is therefore highly probable, that our sun is one of a great number that are in similar circumstances. To this may be added, that the stars we consider as insulated are also surrounded by a magnificent collection of innumerable stars, called the milky-way, which must occasion a very powerful balance of opposite attractions, to hold the intermediate stars in a state of rest. For, though our sun, and all the stars we see; may truly be said to be in the plane of the milky-way, yet I am now convinced, by a long inspection and continued examination of it, that the milky-way itself consists of stars very differently scattered from those which are immediately about us. But of this, more will be said on another occasion.[3]

From the detached situation of insulated stars, it appears that they are capable of being the centres of extensive planetary systems. Of this we have a convincing proof in our sun, which, according to our classification, is one of these stars. Now, as we enjoy the advantage of being able to view the solar system in all its parts, by means of our telescopes, and are therefore sufficiently acquainted with it, there will be no occasion to enter into a detail of its construction.

The question will now arise, whether every insulated star be a sun like ours, attended with planets, satellites, and numerous comets? And here, as nothing appears against the supposition, we may from analogy admit the probability of it. But, were we to extend this argument to other sidereal constructions, or, still farther, to every star of the heavens, as has been done frequently, I should not only hesitate, but even think that, from what will be said of stars which enter into complicated sidereal systems, the contrary is far more likely to be the case; and that, probably, we can only look for solar systems among insulated stars.[4]

II. *Of Binary sidereal Systems, or double Stars*

The next part in the construction of the heavens, that offers itself to our consideration, is the union of two stars, that are formed together into one system, by the laws of attraction.

If a certain star should be situated at any, perhaps immense, distance behind another, and but very little deviating from the line in which we see the first, we should then have the appearance of a double star. But these stars, being totally unconnected, would not form a binary system. If, on the contrary, two stars

[3] [The occasion never came, the nearest to such an account being article 20 of Herschel's 1814 paper (below, pp. 199–200).]

[4] [For most of his career Herschel was insistent that all stars had planetary systems, but his current work on binary stars has brought home to him the perturbations of one star's planetary system that would result from the presence of the other star.]

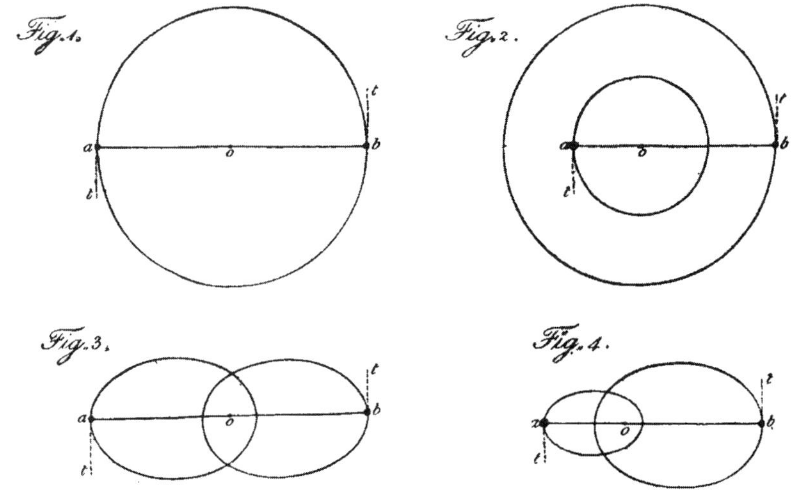

should really be situated very near each other, and at the same time so far insulated as not to be materially affected by the attractions of neighbouring stars, they will then compose a separate system, and remain united by the bond of their own mutual gravitation towards each other. This should be called a real double star; and any two stars that are thus mutually connected, form the binary sidereal system which we are now to consider.

It is easy to prove, from the doctrine of gravitation, that two stars may be so connected together as to perform circles, or similar ellipses, round their common centre of gravity. In this case, they will always move in directions opposite and parallel to each other; and their system, if not destroyed by some foreign cause, will remain permanent.

Figure 1 represents two equal stars *a* and *b*, moving in one common circular orbit round the centre *o*, but in the opposite directions of *a t* and *b t*. In Fig. 2 we have a similar connection of the two stars *a b*; but, as they are of different magnitudes, or contain unequal quantities of matter, they will move in circular orbits of different dimensions round their common centre of gravity *o*. Fig. 3 represents equal, and Fig. 4 unequal stars, moving in similar elliptical orbits round a common centre; and, in all these cases, the directions of the tangents *t t*, in the places *a b*, where the stars are, will be opposite and parallel, as will be more fully explained hereafter.

These four orbits, simple as they are, open an extensive field for reflection, and I may add, for calculation. They shew, even before we come to more complicated combinations, where the same will be confirmed, that there is an essential difference between the construction of solar and sidereal systems. In each solar system, we have a very ponderous attractive centre, by which all the planets, satellites, and comets are governed, and kept in their orbits. Sidereal systems take a greater scope: the stars of which they are composed move round an empty centre, to which they are nevertheless as firmly bound as the planets to their massy one. It is however not necessary here to enlarge on distinctions which will hereafter be strongly support by facts, when clusters of stars come to be considered. I shall only add, that in the subordinate bodies of the solar system itself, we have already instances, in miniature, as it may be called, of the principle whereby the laws of attraction are applicable to the solution of the most complicated phenomena of the heavens, by means of revolutions round empty centres. For, although both the earth and its moon are retained in their orbits by the sun, yet their mutual subordinate system is such that they perform secondary monthly revolutions round a centre without a body placed in it. The same indeed, though under very narrow limits, may be said of the sun and each planet itself.

That no insulated stars, of nearly an equal size and distance, can appear double to us, may be proved thus. Let Arcturus and Lyra [Vega] be the stars: these, by the rule of insulation, which we must now suppose can only take place when their distant from each other is not less than that of Sirius from us, if very accurately placed, would be seen under an angle of 60 degrees from each other. They really are about 59°. Now, in order to make these stars appear to us near enough to come under the denomination of a double star of the first class, we should remove the earth from them at least 41253 times farther than Sirius is from us. But the space-penetrating power of a 7-feet reflector, by which my observations on double stars have been made, cannot intitle us to see stars at such an immense distance; for even the 40-feet telescope, as has been shewn,[5] can only reach stars of the 1342d magnitude. It follows, therefore, that these stars could not remain visible in a 7-feet reflector, if they were so far removed as to make their angular distance less than about 24¼ minutes; nor could even the 40-feet telescope, under the same circumstances of removal, shew them, unless they were to be seen at least 2½ minutes asunder. Moreover, this calculation is made on a supposition that the stars of which a double star is composed, might be as small as any that can possibly be perceived; but if, on the contrary, they should still appear of a considerable size, it will then be so much the more evident that such stars cannot have any great real distance, and that, consequently, insulated stars cannot appear double, if they are situated at equal distances from us. If, however, their arrangement should be such as has been mentioned before, then, one of them being far behind the other, an apparent double star may certainly be produced; but here the appearance of proximity

[5] See Phil. Trans. for 1800, Part I, page 83.

would be deceptive; and the object so circumstanced could not be classed in the list of binary systems. However, as we must grant, that in particular situations stars apparently double may be composed of such as are insulated, it cannot be improper to consult calculation, in order to see whether it be likely that the 700 double stars I have given in two catalogues, as well as many more I have since collected, should be of that kind. Such an inquiry, though not very material to our present purpose, will hereafter be of use to us, when we come to consider more complicated systems. For, if it can be shown that the odds are very much against the casual production of double stars, the same argument will be still more forcible, when applied to treble, quadruple, or multiple compositions.

Let us take ζ Aquarii, for an instance of computation. This star is admitted by Flamsteed, De la Caille, Bradley, and Mayer, to be of the 4th magnitude. The two stars that compose it being equal in brightness, each of them may be supposed to shine with half the light of the whole lustre. This, according to our way of reckoning magnitudes,[6] would make them $4m \times \sqrt{2} = 5\tfrac{2}{3}m$; that is, of between the 6th and 5th magnitude each. Now, the light we receive from a star being as the square of its diameter directly, and as the square of its distance inversely, if one of the stars of ζ Aquarii be farther off than the stars of between the 6th and 5th magnitude are from us, it must be so much larger in diameter, in order give us an equal quantity of light. Let it be at the distance of the stars of the 7th magnitude; then its diameter will be to the diameter of the star which is nearest to us as 7 to 5⅔, and its bulk as 1,885 to 1; which is almost double that of the nearest star. Then, putting the number of stars we call of between the 6th and 5th magnitude at 450, we shall have 686 of the 7th magnitude to combine with them, so that they may make up a double star of the first class, that is to say, that the two stars may not be more than 5″ asunder. The surface of the globe contains 34036131547 circular spaces, each of 5″ in diameter; so that each of the 686 stars will have 49615357 of these circles in which it might be placed; but, of all that number, a single one would only be the proper situation in which it could make up a double star with one of the 450 given stars. But these odds, which are above 75½ millions to one against the composition of ζ Aquarii, are extremely increased by our foregoing calculation of the required size of the star, which must contain nearly double the mass allotted to other stars of the 7th magnitude; of which, therefore, none but this one can be proper for making up the required double star. If the stars of the 8th and 9th magnitudes, of which there will be 896 and 1134, should be taken in, by way of increasing the chance in favour of the supposed composition of our double star, the advantage intended to be obtained

[6] The expressions 2m, 3m, 4m, &c. stand for stars at the distance of 2, 3, 4, &c. times that of Sirius, supposed unity.

by the addition of numbers, will be completely counteracted by the requisite uncommon bulk of the star which is to serve the purpose; for, one of the 8th magnitude ought to be more than 2¾ times bigger than the rest; and, if the composition were made by a star of the 9th magnitude, no less than four times the bulk of the other star which is to enter the composition of the double star would answer the purpose of its required brightness. Hence therefore it is evident, that casual situations will not account for the multiplied phenomena of double stars, and that consequently their existence must be owing to the influence of some general law of nature; now, as the mutual gravitation of bodies towards each other is quite sufficient to account for the union of two stars, we are authorised to ascribe such combinations to that principle.

It will not be necessary to insist any further on arguments drawn from calculation, as I shall soon communicate a series of observations made on double stars, whereby it will be seen, that *many of them have actually changed their situation with regard to each other, in a progressive course, denoting a periodical revolution round each other; and that the motion of some of them is direct, while that of others is retrograde.* Should these observations be found sufficiently conclusive, we may already have their periodical times near enough to calculate, within a certain degree of approximation, the parallax and mutual distance of the stars which compose these systems, by measuring their orbits, which subtend a visible angle.

Before we leave the subject of binary systems, I should remark, that it evidently appears, that our sun does not enter into a combination with any other star, so as to form one of these systems with it. This could not take place without our immediately perceiving it; and, though we may have good reason to believe that our system is not perfectly at rest, yet the causes of its proper motion are more probably to be ascribed to some perturbations arising from the proper motion of neighbouring stars or systems, than to be placed to the account of a periodical revolution round some imaginary distant centre.

III. Of more complicated sidereal Systems, or treble, quadruple, quintuple, and multiple Stars

IV. Of clustering Stars, and the Milky-way

From quadruple, quintuple, and multiple stars, we are naturally led to a consideration of the vast collections of small stars that are profusely scattered over the milky-way. On a very slight examination, it will appear that this immense starry aggregation is by no means uniform. The stars of which it is composed are very unequally scattered, and shew evident marks of clustering together into many separate allotments. By referring to some one of these clustering collections in the heavens, what will be said of them will be much better understood, than if we were to treat of them merely in a general way. Let us take the space between β and γ Cygni for an example, in which the stars are clustering with a kind of division between them, so that we may suppose them to be clustering towards two different regions. By a computation, founded on observations which ascertain the number of stars in different fields of view, it appears that our space between β and γ, taking an average breadth of about five degrees of it, contains more than 331 thousand stars; and, admitting them to be clustering two different ways, we have 165 thousand for each clustering collection. Now, as a more particular account of the milky-way will be the subject of a separate paper, I shall only observe, that the above mentioned milky appearances deserve the name of clustering collections, as they are certainly brighter about the middle, and fainter near their undefined borders. For, in my sweeps of the heavens, it has been fully ascertained, that the brightness of the milky-way arises only from stars; and that their compression increases in proportion to the brightness of the milky-way.

We may indeed partly ascribe the increase, both of brightness and of apparent compression, to a greater depth of the space which contains these stars; but this will equally tend to shew their clustering condition: for, since the increase of brightness is gradual, the space containing the clustering stars must tend to a spherical form, if the gradual increase of brightness is to be explained by the situation of the stars.

V. *Of Groups of Stars*

From clustering stars there is but a short transition to groups of stars; they are, however, sufficiently distinct to deserve a separate notice. A group is a collection of closely, and almost equally compressed stars, of any figure or outline; it contains no particular condensation that might point out the seat of an hypothetical central force; and is sufficiently separated from neighbouring stars to shew that it makes a peculiar system of its own. It must be remembered, that its being a separate system does not exclude it from the action or influence of other systems. We are to understand this with the same reserve that has been pointed out, when we explained what we called insulated stars.

The construction of groups of stars is perhaps, of all the objects in the heavens, the most difficult to explain; much less can we now enter into a detail of the numerous observations I have already made upon this object. I therefore proceed in my enumeration.

VI. *Of Clusters of Stars*

These are certainly the most magnificent objects that can be seen in the heavens. They are totally different from mere groups of stars, in their beautiful and artificial arrangement: their form is generally round; and the compression of the stars shews a gradual, and pretty sudden accumulation towards the centre, where, aided by the depth of the cluster, which we can have no doubt is of a globular form, the condensation is such, that the stars are sufficiently compressed to produce a mottled lustre, nearly amounting to the semblance of a nucleus. A centre of attraction is so strongly indicated, by all the circumstances of the appearance of the cluster, that we cannot doubt a single moment of its existence, either in a state of real solidity, or in that of an empty centre, possessed of an hypothetical force, arising from the joint exertion of the numerous stars that enter into the composition of the cluster.

The number of observations I have to give relating to this article, in which my telescopes, especially those of high space-penetrating power, have been of the greatest service, of course can find no room in this enumeration.

VII. *Of Nebulae*

These curious objects, which, on account of their great distance, can only be seen by instruments of great space-penetrating power, are perhaps all to be resolved into the three last mentioned species. Clustering collections of stars, for instance, may easily be supposed sufficiently removed to present us with the appearance of a nebula of any shape, which, like the real object of which it is the miniature, will seem to be gradually brighter in the middle. Groups of stars also may, by distance, assume the semblance of nebulous patches; and real clusters of stars, for the same reason, when their composition is beyond the reach of our most powerful instruments to resolve them, will appear like round nebula that are gradually much brighter in the middle. On this occasion I must remark, that with instruments of high space-penetrating powers, such as my 40-feet telescope, nebulae are the objects that may be perceived at the greatest distance. Clustering collections of stars, much less than those we have mentioned before, may easily contain 50000 of them; and, as that number has been chosen for an instance of calculating the distance at which one of the most remote objects might be still visible,[7] I shall take notice of an evident consequence attending the result of the computation; which is, that a telescope with a power of penetrating into space, like my 40-feet one, has also, as it may be called, a power of penetrating into time past. To explain this, we must consider that, from the known velocity of light, it may be proved, that when we look at Sirius, the rays which enter the eye cannot have been less than 6 years and 4½ months coming from that star to the observer. Hence it follows, that when we see an object of the calculated distance at which one of these very remote nebulae may

[7] See Phil. Trans. for 1800, page 83.

still be perceived, the rays of light which convey its image to the eye, must have been more than nineteen hundred and ten thousand, that is, almost two millions of years on their way; and that, consequently, so many years ago, this object must already have had an existence in the sidereal heavens, in order to send out those rays by which we now perceive it.

VIII. *Of Stars with Burs, or Stellar Nebulae*

Situated as we are, at an immense distance from the remote parts of the heavens, it is not in the power of telescopes to resolve many phenomena we can but just perceive, which, could we have a nearer view of them, might probably shew themselves as objects that have long been known to us. A stellar nebula, perhaps, may be a real cluster of stars, the whole light of which is gathered so nearly into one point, as to leave but just enough of the light of the cluster visible to produce the appearance of burs. This, however, admits of a doubt.

IX. *Of milky Nebulosity*

The phenomenon of milky nebulosity is certainly of a most interesting nature: it is probably of two different kinds; one of them being deceptive, namely, such as arises from widely extended regions of closely connected clustering stars, contiguous to each other, like the collections that construct our milky-way. The other, on the contrary, being real, and possibly at no very great distance from us. The changes I have observed in the great milky nebulosity of Orion, 23 years ago, and which have also been noticed by other astronomers, cannot permit us to look upon this phenomenon as arising from immensely distant regions of fixed stars. Even HUYGENS, the discoverer of it, was already of opinion that, in viewing it, we saw, as it were, through an opening into a region of light.[8] Much more would he be convinced now, when changes in its shape and lustre have been seen, that its light is not, like that of the milky-way, composed of stars. To attempt even a guess at what this light may be, would be presumptuous. If it should be surmised, for instance, that this nebulosity is of the nature of the zodiacal light, we should then be obliged to admit the existence of an effect without its cause. An idea of its phosphorical condition, is not more philosophical, unless we could shew from what source of phosphorical matter, such immeasurable tracts of luminous phenomena could draw their existence, and permanency; for, though minute changes have been observed, yet a general resemblance, allowing for the difference of telescopes, is still to be perceived in the great nebulosity of Orion, even since the time of its first discovery.

[8] See *Systema Saturnium*, page 8 and 9.

X. Of nebulous Stars

The nature of these remarkable objects is enveloped in much obscurity. It will probably require ages of observations, before we can be enabled to form a proper estimate of their condition. That stars should have visible atmospheres, of such an extent as those of which I have given the situation in this and my former catalogues, is truly surprising, unless we attribute to such atmospheres, the quality of self-luminous milky nebulosity. We can have no reason to doubt of the starry nature of the central point; for, in no respect whatever does its appearance differ from that of a star of an equal magnitude; but, when the great distance of such stars is taken into consideration, the real extent of the surrounding nebulosity is truly wonderful. A very curious one of this kind will be found in the 4th class, No. 69, of the annexed catalogue.[9]

XI. Planetary Nebulae

This seems to be a species of bodies that demands a particular attention. To investigate the planetary nature of these nebulae, is not an easy undertaking. If we admit them to contain a great mass of matter, such as that of which our sun is composed, and that they are, like the sun, surrounded by dense luminous clouds, it appears evidently that the intrinsic brightness of these clouds must be far inferior to those of the sun. A part of the sun's disk, equal to a circle of 15″ in diameter, would far exceed the greatest lustre of the full moon; whereas, the light of a planetary nebula, of an equal size, is hardly equal to that of a star of the 8th or 9th magnitude. If, on the other hand, we should suppose them to be groups, or clusters of stars, at a distance sufficiently great to reduce them to so small an apparent diameter, we shall be at a loss to account for their uniform light, if clusters; or for their circular forms, if mere groups of stars.

Perhaps they may be rather allied to nebulous stars. For, should the planetary nebulae with lucid centres, of which the next article will give an account, be an intermediate step between planetary nebulae and nebulous stars, the appearances of these different species, when all the individuals of them are fully examined, might throw a considerable light upon the subject.

XII. Of planetary Nebulae with Centres

In my second catalogue of nebulae, a single instance of a planetary nebula with a bright central point was mentioned; and, in the annexed one, No. 73 of the 4th class, is another of very nearly the same diameter, which has

[9] [The planetary nebula NGC 1514, see pp. 152–153 above. Note that for Herschel, the visible central star makes it for him a nebulous star rather than a planetary nebula.]

also a lucid, though not quite so regular a centre.[10] From several particularities observed in their construction, it would seem as if they were related to nebulous stars. If we might suppose that a gradual condensation of the nebulosity about a nebulous star could take place, this would be one of them, in a very advanced state of compression. A further discussion of this point, however, must be reserved to a future opportunity.

[William Herschel.]

[10] [The planetary nebulae NGC 40 and NGC 6826. – S.]

1811: "Observations relating to the construction of the heavens"

Astronomical Observations relating to the Construction of the Heavens, arranged for the Purpose of a critical Examination, the Result of which appears to throw some new Light upon the Organization of the celestial Bodies[1]

Read June 20, 1811.

A KNOWLEDGE of the construction of the heavens has always been the ultimate object of my observations, and having been many years engaged in applying my forty, twenty, and large ten feet telescopes, on account of their great space-penetrating power to review the most interesting objects discovered in my sweeps, as well as those which had before been communicated to the public in the *Connaissance des Temps*, for 1784, I find that by arranging these objects in acertain successive regular order, they may be viewed in a new light, and, if I am not mistaken, an examination of them will lead to consequences which cannot be indifferent to an inquiring mind.[2]

If it should be remarked that in this new arrangement I am not entirely consistent with what I have already in former papers said on the nature of some objects that have come under my observation, I must freely confess that by continuing my sweeps of the heavens my opinion of the arrangement of the stars and their magnitudes, and of some other particulars, has undergone a gradual change; and

[1] [*Phil. Trans.*, **101** (1811), 269–336.]

[2] [This paper and its companion together occupy more than a hundred pages in the original. Fortunately this length is due to Herschel's desire to reduce to a minimum the step from one 'article' to the next and at the same time to multiply his examples, so that the substance of the papers can be reproduced in a shorter space. Thus for some intermediate 'articles' the title alone conveys Herschel's meaning.]

indeed when the novelty of the subject is considered, we cannot be surprised that many things formerly taken for granted, should on examination prove to be different from what they were generally, but incautiously, supposed to be. For instance, an equal scattering of the stars may be admitted in certain calculations; but when we examine the milky way, or the closely compressed clusters of stars, of which my catalogues have recorded so many instances, this supposed equality of scattering must be given up. We may also have surmised nebulae to be no other than clusters of stars disguised by their very great distance, but a longer experience and better acquaintance with the nature of nebulae, will not allow a general admission of such a principle, although undoubtedly a cluster of stars may assume a nebulous appearance when it is too remote for us to discern the stars of which it is composed.[3]

Impressed with an idea that nebulae properly speaking were clusters of stars, I used to call the nebulosity of which some were composed, when it was of a certain appearance, *resolvable*; but when I perceived that additional light, so far from resolving these nebulae into stars, seemed to prove that their nebulosity was not different from what I had called milky, this conception was set aside as erroneous. In consequence of this, such nebulae as afterwards were suspected to consist of stars, or in which a few might be seen, were called *easily resolvable*; but even this expression must be received with caution, because an object may not only contain stars, but also nebulosity not composed of them. It will be necessary to explain the spirit of the method of arranging the observed astronomical objects under consideration in such a manner, that one shall assist us to understand the nature and construction of the other. This end I propose to obtain by assorting them into as many classes as will be required to produce the most gradual affinity between the individuals contained in any one class with those contained in that which precedes and that which follows it: and it will certainly contribute to the perfection of this method, if this connection between the various classes can be made to appear so clearly as not to admit of a doubt. This consideration will be a sufficient apology for the great number of assortments into which I have thrown the objects under consideration; and it will be found that those contained in one article, are so closely allied to those in the next, that there is perhaps not so much difference between them, if I may use the comparison, as there would be in an annual description of the human figure, were it given from the birth of a child till he comes to be a man in his prime.

The similarity of the objects contained in each class will seldom require the description of more than one of them, and for this purpose, out of the number referred to, the selected one will be that which has been most circumstantially

[3] [A frank admission of a valuable assumption abandoned and a major reappraisal undertaken.]

observed; however, those who wish either to review any other of the objects, or to read a short description of them, will find their place in the heavens, or the account of their appearance either in the catalogues I have given of them in the Philos. Trans. or in the *Connoissance des Temps* for 1784, to which in every article proper references will be given for the objects under consideration.

If the description I give should sometimes differ a little from that which belongs to some number referred to, it must be remembered that objects which had been observed many times, could not be so particularly and comprehensively detailed in the confined space of the catalogues as I now may describe them; additional observations have also now and then given me a better view of the objects than I had before. This remark will always apply to the numbers which refer to the *Connoissance des Temps*; for the nebulae and clusters of stars are there so imperfectly described, that my own observation of them with large instruments may well be supposed to differ entirely from what is said of them. But if any astronomer should review them, with such high space-penetrating-powers, as are absolutely required, it will be found that I have classed them very properly.

It will be necessary to mention that the nebulous delineations in the figures are not intended to represent any of the individuals of the objects which are described otherwise than in the circumstances which are common to the nebulae of each assortment: the irregularity of a figure, for instance, must stand for every other irregularity; and the delineated size for every other size.[4] It will however be seen, that in the figure referred to there is a sufficient resemblance to the described nebula to show the essential features of shape and brightness then under consideration.

1. Of extensive diffused Nebulosity

The first article of my series will begin with extensive diffused nebulosity, which is a phenomenon that hitherto has not been much noticed, and can indeed only be perceived by instruments that collect a great quantity of light. Its existence, when some part of it is pointed out by objects that are within the reach of common telescopes, has nevertheless obtruded itself already on the knowledge of astronomers, as will be seen in my third article.

The widely diffused nebulosity under consideration has already been partially mentioned in my catalogues.[5]

[4] [Herschel's cavalier attitude to the depiction of nebulae contrasts with that of his son John who went to great pains to represent an object in full detail. – S.]

[5] See *Phil. Trans.* 1786, p. 471; 1789, p. 226; and 1802, p. 503. The following ten nebulosities are in the Vth class, No. 13, 14, 15, 17, 28, 30, 31, 33, 34, 38. [H. V. 33 = NGC 1908 was observed only once by Herschel (1 Feb. 1786), but as there is no diffused nebulosity in the place he cites (SW of η Orionis) the object has the status "not found". – S.]

The description of the object I shall select is of No. 14 in the 5th class,[6] and is as follows: "Extremely faint branching nebulosity; its whitishness is entirely of the milky kind, and it is brighter in three or four places than in the rest; the stars of the milky way are scattered over it in the same manner as over the rest of the heavens. Its extent in the parallel is nearly 1½ degree, and in the meridional direction about 52 minutes. The following part of it is divided into several streams and windings, which after separating, meet each other again towards the south." See Fig. 1.

This account, which agrees with what will be found in all the other numbers referred to, with regard to the subject under consideration, namely, a diffused milky nebulosity, will give us already some idea of its great abundance in the heavens; my next article however will far extend our conception of its quantity.

2. Observations of Nebulosities that have not been published before[7]

3. Of Nebulosities joined to Nebulae

4. Of detached Nebulosities

5. Of milky Nebulae

When detached nebulosities are small we are used to call them nebulae, and it is already known from my catalogues that their number is very great. . ..[8]

[6] [H. V. 14 = NGC 6992, the f. part of the Cygnus Loop system. Herschel has previously used this as an example of a "Milky Way" in his 1785 paper (p. 130), supposing it to be made of stars: he now regards it, correctly, as true nebulosity. – D.]

[7] [This article 2 is a discussion and list of fifty-two fields "affected with nebulosity". These were the subject of considerable controversy in the early decades of the last century. Many of the fields are in high galactic latitude where little gaseous nebulosity would be expected: some are near the ecliptic and Herschel may have been misled by the zodiacal light. – D.] [There are only seven regions that can be linked to existing nebulosities with some confidence: nos. 22–25 and 27 are located in the Orion region (no. 27 may be the brightest part of Barnard's Loop). Nos. 42 and 44 are in Cygnus and may be partly identical with the North American Nebula, NGC 7000. The problem is, how could Herschel perceive these extended features when his field of view covered a mere 15′? Visually there is no contrast whatever if one looks inside such a region. Only the "edges" might be noticeable, although these are scarcely perceptible. – S.]

[8] [An important definition of terms. ". . . their number is very great": most (but not all) of the nebulae referred to in this paper we should now call galaxies. Although he discovered and gave catalogue numbers to many gaseous nebulae in the Milky Way, most of Herschel's new nebulae are in fact the galaxies in high galactic latitude whose existence in large numbers he was the first to describe. – D.]

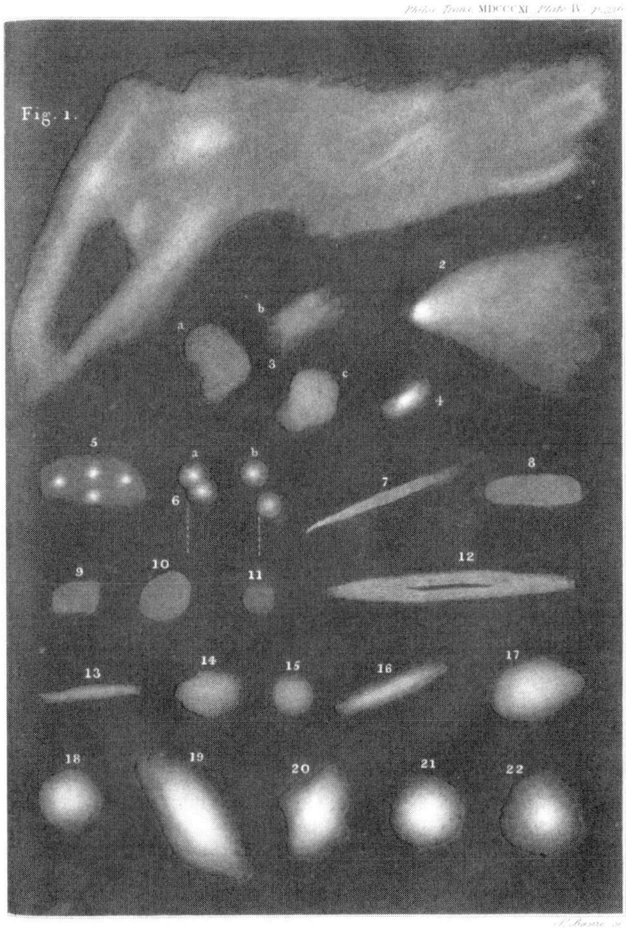

6. Of milky Nebulae with Condensation

7. Of Nebulae which are brighter in more than one Place

It is not an uncommon circumstance that the same nebula is brighter in several different places than in the rest of its compass. The following six are of this sort.[9]

No. 213 in the first class[10] is "A very brilliant and considerably large nebula, extended in a direction from south preceding to north following. It seems to have three or four bright nuclei." See Fig. 5.

[9] See I. 165, 213, 261. II. 297, 406. III. 49.

[10] [H. I. 213 = NGC 4449. A bright galaxy, type Irr I, elongated, with star clouds and patches

From this construction of the nebulae, we may draw some additional information concerning the point which was left undetermined in my last article; for since there it was proposed as an alternative, that the nebulous matter might either be of a greater depth or more compressed in the brightest part of the nebula then under consideration, we have now an opportunity to examine the probability of each case. If here the appearance of several bright nuclei is to be explained by the depth of the nebulous matter, we must have recourse to three or four separate very slender and deep projections, all situated exactly in the line of sight; but such a very uncommon arrangement of nebulous matter cannot pretend to probability; whereas a moderate condensation, which may indeed be also accompanied with some little general swelling of the nebulous matter about the places which appear like nuclei, will satisfactorily account for their superior brightness.

The same method of reasoning may be as successfully applied to explain the number of unequally bright places in the diffused nebulosities which have been described in the 1st, 2d, and 3d articles. For instance, in the branching nebulosity V. 14,[11] we find three or four places brighter than the rest – in the nebulosity No. 44 of the table[12] we have places of different brightness. In the nebula of Orion, there are many parts that differ much in lustre; and in V. 37[13] of the same article I found, by an observation in the year 1790, the same variety of appearance. In all these cases a proportional condensation of the nebulous matter in the brighter places will sufficiently account for their different degree of shining.

This way of explaining the observed appearances being admitted, it will be proper to enter into an examination of the probable cause of the condensation of the nebulous matter. Should the necessity for such a condensing cause be thought to be admitted upon too slight an induction, a more detailed support of it will hereafter be found in the condition of such a copious collection of objects, as will establish its existence beyond all possibility of doubt.[14]

Instead of inquiring after the nature of the cause of the condensation of nebulous matter, it would indeed be sufficient for the present purpose to call it

of emission nebulosity. – D.] [Visually there are many bright H II regions in this "magellanic" type object, of which Herschel sketched the four most striking. – S.]

11 [H. V. 14 = NGC 6992. Veil Nebula of Cygnus Loop system. – D.]

12 ["the nebulosity No. 44 of the table": i.e. no. 44 in the table of fifty-two regions given, but not here reproduced, in Article 2 above. This area is NGC 7000, the North America nebula. – D.]

13 [H. V. 37 = NGC 7000, again. For the complex history of Herschel's observations of this object, see Wolfgang Steinicke, *Observing and Cataloguing Nebulae and Star Clusters* (Cambridge, 2010), 32–33. – D.]

14 See Article 24.

merely a condensing principle; but since we are already acquainted with the centripetal force of attraction which gives a globular figure to planets, keeps them from flying out of their orbits in tangents, and makes one star revolve around another, why should we not look up to the universal gravitation of matter as the cause of every condensation, accumulation, compression, and concentration of the nebulous matter? Facts are not wanting to prove that such a power has been exerted; and as I shall point out a series of phenomena in the heavens where astronomers may read in legible characters the manifest vestiges of such an exertion, I need not hesitate to proceed in a few additional remarks on the consequences that must arise from the admission of this attractive principle.

The nebula, for instance, which has been described at the beginning of this article, as containing several bright nuclei, has probably so many predominant seats of attraction, arising from a superior preponderance of the nebulous matter in those places; but attraction being a principle which never ceases to act, the consequence of its continual exertion upon this nebula will probably be a division of it, from which will arise three or four distinct nebulae. In the same manner its operation on the diffused nebulosities that have many different bright places, will possibly occasion a breaking up of them into smaller diffusions and detached nebulae; but before I proceed with conjectures, let us see what observations we have to give countenance to such expectations.

8. Of double Nebulae with joined Nebulosity

9. Of double Nebulae that are not more than two Minutes from each other

To add to the probability of the separation of nebulae, we ought to have a considerable number of them already separated. The following twenty-three are completely divided although not more than two minutes from one another.[15]

A description of II. 714[16] is "Two pretty bright nebulae; they are both round, small, and about 2′ from each other, in a meridional direction."

Of III. 755[17] is "Two very faint, very small extended nebulae within 1½′ from each other."

[15] See I. 116, 190, 197. II. 8, 28, 57, 111, 178, 450, 714. III. 92, 228, 280, 591, 687, 719, 755, 855, 943, 952, 959, 967.

[16] [H. II. 714 = NGC 5353. Two close galaxies, E and S0, of similar magnitude. – D.]

[17] [H. III. 755 = NGC 4403–4. Two close galaxies, E and S, of similar magnitude: the S galaxy shows evidence of interaction indicating physical proximity. – D.]

That all these nebulae are really double, is founded on the reason already assigned in the last article.[18] Then if we would enter into some kind of examination how they came to be arranged into their binary order, we cannot have recourse to a promiscuous scattering, which by a calculation of chances can never account for such a peculiar distribution of them.[19] If, on the contrary, we look to a division of nebulous matter by the condensing principle, then every parcel of it, which had more than one preponderating seat of attraction in its extent, must in the progress of time have been divided.

No doubt can be suggested on account of the great length of time such a division must have taken up, when we have an eternity of past duration to recur to.

10. Of double Nebulae at a greater Distance than 2′ from each other

11. Of treble, quadruple, and sextuple Nebulae[20]

12. Of the remarkable Situation of Nebulae

The number of compound[21] nebulae that have been noticed in the foregoing three articles being so considerable, it will follow, that if they owe

[18] [Most of the twenty-three objects are close pairs of galaxies, some in physical proximity. A few of the nebulae listed as double nebulae are in fact relatively large single galaxies in which Herschel has seen two brighter separate condensations of light and listed them as separate objects in his catalogue. Different parts of a single large galaxy may bear different NGC numbers for this reason. – D.]

[19] [Herschel gladly adapts the earlier probability arguments in favour of binary systems of stars, above, p. 18.]

[20] [Herschel lists twenty examples and describes three in detail. They are mostly closely-grouped galaxies. The first described is however H. V. 10 = NGC 6514, the Trifid Nebula (gaseous). The other two are H. III. 358 = NGC 4169, 4173, 4174, 4175, four galaxies in a close group; and H. III. 391 = NGC 4070, . . ., six galaxies, the brightest at the n.f end of a scattered cluster. – D.] [The Trifid Nebula, H. V. 10, is identical with V. 11, V. 12, and IV. 41, the multiple identities being due to a position error by Herschel. – S.]

[21] ["The number of compound nebulae . . . noticed in the foregoing three articles." Compound here apparently used in the sense of double, triple or multiple, rather than, in the sense (compound nebula = a Milky Way) of p. 119. Although a few gaseous nebulae like NGC 6514 (the Trifid) have been included, most of the nebulae referred to are close pairs or groups of galaxies.

The substance of this Article 12 is to draw attention to large numbers of nebulae, almost all galaxies, and remark upon their concentration in certain areas of sky, especially in the constellations of Virgo, Coma, etc., with subsidiary concentrations in three areas ("branches") in Bootes, Ursa Major, and Leo. The areas where "the absence of nebula is as remarkable . . ." are for the greater part in lower galactic latitude or free from prominent groupings of galaxies like those near the north galactic pole. – D.]

their origin to the breaking up of some former extensive nebulosities of the same nature with those which have been shewn to exist at present, we might expect that the number of separate nebulae should far exceed the former, and that moreover these scattered nebulae should be found not only in great abundance, but also in proximity or continuity with each other, according to the different extents and situations of the former diffusions of such nebulous matter. Now this is exactly what by observation, we find to be the state of the heavens.

In the following seven assortments we have not less than 424 nebulae; some of them of unascertained size, figure, or condensation; and the rest with only the first of these three essential features recorded.

The reason for not having a more circumstantial account of such a number of objects, is that they crowded upon me at the time of sweeping in such quick succession, that of sixty-one I could but just secure the place in the heavens, and of the remaining three hundred and sixty-three, I had only time to add the relative size.[22]

Neither of the nebulae in these seven divisions will require a description, as the title of each assortment contains all that has been ascertained about them; but their number and situation, especially when added to those that will be contained in the following articles, completely supports what has been asserted, namely, that the present state of the heavens presents us with several extensive collections of scattered nebulae, plainly indicating by their very remarkable arrangement, that they owe their origin to some former common stock of nebulous matter.

To refer astronomers to the heavens for an inspection of these and the following nebulae, would be to propose a repetition of more than eleven hundred sweeps to them, but those who wish to have some idea of the nebulous arrangements may consult Mr. Bode's excellent Atlas Coelestis.[23] A succession of places where the nebulae of my catalogues are uncommonly crowded, will there be seen beginning over the tail of Hydra and proceeding to the southern wing, the body and the northern wing of Virgo, Plate 14. Then to Coma Berenices, Canes venatici,

[22] See *sixty-one nebulae. ... Ten extremely small nebulae. ... One hundred and thirty-six very small nebulae. ... Forty-two not very small nebulae. ... One hundred and seven small nebulae. ... Fifty-eight pretty large nebulae. ... Ten large nebulae. ...* [Catalogue numbers not reproduced.]

[23] [J. E. Bode, *Uranographia* (Berlin, 1801). In the "Praefatio" of this magnificent collection of sky charts, Bode refers to "Celeberrimus Herschelius", "the most famous Herschel", and among his (newly-invented) constellations is Telescopium Herschelii. Each nebula (*Nebulositas*) and star cluster (*Conglobatio*) of Herschel's catalogues is marked along with its classification, and each double star (*Duplicitas*) is also shown.]

and the preceding arm of Bootes, Plate 7. A different branch goes from Coma Berenices to the hind legs of Ursa major. Another branch passes from the wing of Virgo to the tail and body of Leo, Plate 8.

It will not be necessary to point out many other smaller collections which may be found in several plates of the same Atlas.

On the other hand, a very different aspect of the heavens will be perceived when we examine the following constellations. Beginning from the head of Capricorn, Plate 16, thence proceeding to Antinous, to the tail of Aquila, Plate 9, to Ramus Cerberus, and the body of Hercules, Plate 8, to Quadrans Muralis, Plate 7, and to the head of Draco, Plate 3. We may also examine the constellations of Auriga, Lynx, and Camelopardalus, Plate 5.

In this second review, it will be found that here the absence of nebulae is as remarkable, as the great multitude of them in the first mentioned series of constellations.

13. Of very narrow long Nebulae

14. Of extended Nebulae

15. Of Nebulae that are of an irregular Figure

16. Of Nebulae that are of an irregular round Figure

17. Of round Nebulae

18. Of Nebulae that are remarkable for some particularity in Figure or Brightness

19. Of Nebulae that are gradually a little brighter in the middle[24]

The investigation of the form of the nebulous matter in the 13, 14, 15, and 16th articles has been founded only upon the observed figure of nebulae; and in the 17th article the globular form of this matter deduced from the round appearance of nebulae, has been ascribed to the action of the gravitating principle. I am now entering upon an examination of nebulae of which, besides their figure, I have also recorded the differerent degrees of light, and the situation of the greatest brightness with respect to their figure. These observations will establish the former conclusions by an additional number of objects, and by the decisive argument of their brightness, which points out a seat of attraction.

[24] [The nebulae referred to in Articles 13–19, several hundred in number, have not been examined individually; but the type objects described in detail are all galaxies. – D.]

In the following four assortments are one hundred and fifty nebulae, which all agree in being a little brighter in the middle. This increase of brightness must be understood to be always very gradual from the outside towards the middle of the nebula, whatever be its figure; and although this circumstance, for want of time, has often been left unnoticed in the observation, I am very sure that had the gradation of brightness been otherwise, it would certainly not have been overlooked.[25]

III. 853 is "A very faint small nebula; it is very gradually a little brighter in the middle."[26]

III. 488 is "A very faint extended nebula, near 3′ long, and above 2′ broad; it is gradually a little brighter in the middle." Fig. 13.[27]

II. 549 is "A very large and pretty bright nebula of an irregular figure; it is a little brighter in the middle." Fig. 14.[28]

II. 812 is "A faint, small, round nebula; it is very gradually a little brighter in the middle, and the increase of brightness begins at a distance from the center." Fig. 15.[29]

It is hardly necessary to say that the united testimony of so many objects can leave no doubt about the central seat of attraction, which in every instance of figure is pointed out to be in the middle.

The only remark I have to make, relates to the exertion of the condensing power, which in the case of these nebulae appears to have produced but a very moderate effect. This may be ascribed either to the unshapen mass of nebulous matter which would require much time before it could come to some central arrangement of form either in length, or in length and breadth, or lastly in all its three dimensions. It may also be ascribed to the small quantity of the preponderating central attractive matter; or even to the shortness of its time of acting: for in this case millions of years, perhaps are but moments.

20. Of Nebulae which are gradually brighter in the middle

By the general description of a nebula, when it is said to be gradually brighter in the middle, we are to understand that its light was observed to be

[25] See *thirty-two nebulae, the particular figure of which has not been ascertained, gradually a little brighter in the middle. ... Twenty-four extended nebulae, gradually a little brighter in the middle. ... Twenty nebulae of an irregular figure, gradually a little brighter in the middle. ... Seventy-four round or nearly round nebulae, gradually a little brighter in the middle. ...* [Catalogue numbers not reproduced.]

[26] [H. III. 853 = NGC 3073, E/SB0 galaxy. – S.]

[27] [H. III. 488 = NGC 2848, SBc galaxy. – S.]

[28] [H. II. 549 = NGC 4818, SBa galaxy. – S.]

[29] [H. II. 812 = NGC 6338, S0 galaxy. – S.]

obviously brighter about the center than in other parts. Had the nebulae of this class been only a little brighter, or had they been much brighter in the middle, such additional expressions would certainly have been used; except where time would not allow to be more particular. I have sorted two hundred and twenty-three of these nebulae like the foregoing, according to their figure, into four classes.[30]

II. 409[31] is "A pretty bright and pretty large nebula; it is very gradually brighter in the middle."

I. 55[32] is "A considerably bright, extended nebula about 4′ long and 2′ broad, in a meridional direction; it is gradually brighter in the middle." Fig. 16.

I. 266[33] is "A considerably bright, and pretty large nebula, of an irregular figure; it is gradually brighter in the middle." Fig. 17.

I. 98[34] is "A considerably bright, and pretty large round nebula; it is brighter in the middle, the brightness diminishing very gradually from the center towards the circumference." Fig. 18.

From the account of these nebulae, we find again that all what has been said concerning the seat of the forming and condensing power of the nebulous matter, is abundantly confirmed by observation.

I have only to remark that, the exertion of the gravitating principle in these nebulae, is in a more advanced state than with those of the last article; and that the same conceptions which have already been suggested, namely, the original form of the nebulous matter; its quantity in the seat of the attracting principle; and the length of the time of its action, when properly considered, will sufficiently account for the present state of these nebulae.

21. Of Nebulae that are gradually much brighter in the middle

22. Of Nebulae that have a Cometic appearance

Among the numerous nebulae I have seen, there are many that have the appearance of telescopic comets. The following are of that sort.[35]

[30] See *Thirty-nine nebulae of an unascertained figure, gradually brighter in the middle. . . . Fifty extended nebulae gradually brighter in the middle. . . . Twenty-nine nebulae of an irregular figure, gradually brighter in the middle. . . . One hundred and five round, or nearly round nebulae, gradually brighter in the middle. . . .* [Catalogue numbers not reproduced.]

[31] [H. II. 409 = NGC 4190, irregular galaxy – D.]

[32] [H. I. 55 = NGC 7479, SBb galaxy – D.]

[33] [H. I. 266 = NGC 3206, SBc galaxy – D.]

[34] [H. I. 98 = NGC 5273, S0 galaxy – D.]

[35] See *Seventeen cometic nebulae*. I. 3, 4, 34, 217. II. 6, 15, 33, 59, 104, 153, 154, 241, 315, 404. III. 5, 21. *Connoiss*. 95.

I. 4[36] is "A pretty large cometic nebula of considerable brightness; it is much brighter in the middle, and the very faint chevelure is pretty extensive." Fig. 22.

By the appellation of cometic, it was my intention to express a gradual and strong increase of brightness towards the center of a nebulous object of a round figure; having also a faint chevelure or coma of some extent, beyond the faintest part of the light, gradually decreasing from the center.

It seems that this species of nebulae contains a somewhat greater degree of condensation than that of the round nebulae of the last article, and might perhaps not very improperly have been included in their number. Their great resemblance to telescopic comets, however, is very apt to suggest the idea, that possibly such small telescopic comets as often visit our neighbourhood may be composed of nebulous matter, or may in fact be such highly condensed nebulae.

23. Of Nebulae that are suddenly much brighter in the middle

24. Of round Nebulae increasing gradually in brightness up to a Nucleus in the middle

25. Of Nebulae that have a Nucleus

26. Of extended Nebulae that shew the Progress of Condensation

When the nebulous matter is much extended in length, it appears from the following nebulae, that with those which have a nucleus completely formed, the nebulosity on each side of it is comparatively reduced to a fainter state than it is in nebulae of which the nucleus is apparently still in an incipient state. These faint opposite appendages to the nucleus I have in my observations called branches.

In some nebulae there is also an additional small faint nebulosity of a circular form about the nucleus, and this I have called the chevelure. The following two assortments contain twenty-eight nebulae of this kind.[37]

Number 65 of the *Connoissance*[38] is "A very brilliant nebula extended in the meridian, about 12′ long. It has a bright nucleus, the light of which suddenly diminishes on its border, and two opposite very faint branches." Fig. 29.

[36] [H. I. 4 = NGC 3169, Sa galaxy. – D.]

[37] See *Twenty-three extended nebulae with a nucleus and two opposite faint branches*. I. 9, 13, 15, 27, 32, 75, 130, 160, 163, 187, 188, 195, 223, 228, 230. II. 101, 650, 733. IV. 61. V. 43. *Connoiss.* 65, 83, 98. *Five with a nucleus, chevelure and branches*. I. 194, 205, 210. V. 45. *Connoiss.* 94.

[38] [M65 = NGC 3623, a bright Sc galaxy seen steeply inclined, though not edge-on. – D.]

I. 205[39] is "A very brilliant nebula, 5′ or 6′ long and 3 or 4′ broad; it has a small bright nucleus with a faint chevelure about it, and two opposite very extensive branches." Fig. 30.

The construction of these nebulae is certainly complicated and mysterious, and in our present state of knowledge it would be presumptuous to attempt an explanation of it; we can only form a few distant surmises, which however may lead to the following queries.[40] May not the faintness of the branches arise from a gradual diminution, of the length and density of the nebulous matter

[39] [H. I. 205 = NGC 2841, an inclined Sb galaxy with a very bright nucleus and lens-shaped inner region. – D.]

[40] [These rhetorical queries are a stylistic echo of the queries to Newton's *Opticks*.]

contained in them, occasioned by its gravitation towards the nucleus into which it probably subsides? Are not these faint nebulous branches joining to a nucleus, upon an immense scale, somewhat like what the zodiacal light is to our sun in miniature? Does not the chevelure denote that perhaps some of the nebulous matter still remaining in the branches, before it subsides into the nucleus, begins to take a spherical form, and thus assumes the semblance of a faint chevelure surrounding it in a concentric arrangement? And, if we may venture to extend these queries a little farther – will not the matter of these branches in their gradual fall towards the nucleus, when discharging their substance into the chevelure, produce a kind of vortex or rotatory motion? Must not such an effect take place, unless we suppose, contrary to observation, that one branch is exactly like the other; that both are exactly in a line passing through the center of the nucleus, by way of causing exactly an equal stream of it from each branch to enter the chevelure at opposite sides; and, this not being probable, do we not see some natural cause which may give a rotatory motion to a celestial body in its very formation?

27. Of round Nebulae that shew the Progression of Condensation

28. Of round Nebulae that are of an almost uniform Light

29. Of Nebula that draw progressively towards a Period of final Condensation

In the course of the gradual condensation of the nebulous matter, it may be expected that a time must come when it can no longer be compressed, and the only cause which we may suppose to put an end to the compression is, when the consolidated matter assumes hardness. It remains therefore to be examined, how far my observations will go to ascertain the intensity of its consolidation.

The following two assortments contain seven nebulae, from whose appearance a considerable degree of solidity may be inferred.[41]

IV. 55[42] is "A pretty bright round nebula, almost of an even light throughout approaching to a planetary appearance, but ill defined, and a little fainter on the edges; it is about ¾ or 1 minute in diameter." Fig. 34.

IV. 37[43] is "A very bright planetary disk of about 35″ in diameter, but ill defined on the edges; the center of it is rather more luminous than the rest, and

[41] See *Four nebulae of a planetary appearance*. IV. 55, 60, 68, 78. *Three planetary disks with a bright central point*. II. 268. IV. 37, 73.

[42] [H. IV. 55 = NGC 2537, Sc galaxy. – D.] [Visually this object looks like a round planetary nebula, but it is a peculiar galaxy known as the Bear Paw because of its internal structure. – S.]

[43] [H. IV. 37 = NGC 6543, planetary nebula. – D.]

with long attention a very bright well defined round center becomes visible." Fig. 35.

In these nebulae we have three different indications of the compression of the nebulous matter of which they are composed: their figure, their light, and the small compass into which it is reduced. The round figure is a proof that the nebulous mass is collected into a globular form, which cannot have been effected without a certain degree of condensation. . . .

30. Of Planetary Nebulae

31. Of the Distance of the Nebula in the Constellation of Orion

32. Of Stellar Nebulae

33. Of Stellar Nebulae nearly approaching to the Appearance of Stars

34. Of doubtful Nebulae

35. Concluding Remarks

The total dissimilitude between the appearance of a diffusion of the nebulous matter and of a star, is so striking, that an idea of the conversion of the one into the other can hardly occur to any one who has not before him the result of the critical examination of the nebulous system which has been displayed in this paper. The end I have had in view, by arranging my observations in the order in which they have been placed, has been to shew, that the above mentioned extremes may be connected by such nearly allied intermediate steps, as will make it highly probable that every succeeding state of the nebulous matter is the result of the action of gravitation upon it while in a foregoing one, and by such steps the successive condensation of it has been brought up to the planetary condition. From this the transit to the stellar form, it has been shown, requires but a very small additional compression of the nebulous matter, and several instances have been given which connect the planetary to the stellar appearance.

The faint stellar nebulae have also been well connected with all sorts of faint nebulae of a larger size; and in a number of the smaller sort, their approach to the starry appearance is so advanced, that in my observations of many of them it became doubtful whether they were not stars already.

It must have been noticed, that I have confined myself in every one of the preceding articles to a few remarks upon the appearance of the nebulous matter in the state in which my observations represented it; they seemed to be the natural result of the observations under consideration, and were not given with a view to establish a systematic opinion, such as will admit of complete demonstration. The observations themselves are arranged so conveniently that any astronomer, chemist, or philosopher, after having considered my critical remarks, may form what judgment appears most probable to him. At all events, the subject is of such a nature as cannot fail to attract the notice of every inquisitive mind to a contemplation of the stupendous construction of the heavens; and what I have said may at least serve to throw some new light upon the organization of the celestial bodies.[44]

Postscript

It will be seen that in this paper I have only considered the nebulous part of the construction of the heavens, and have taken a star for the limit of my researches. The rich collection of clusters of stars contained in the 6th, 7th, and

[44] [In the preceding thirty-four 'articles', in which Herschel arranges examples of nebulae principally to show the probability of condensation from "extensive diffused Nebulosity" to "Stellar Nebulae nearly approaching to the Appearance of Stars", he cites altogether several hundred examples of nebulae. Their NGC numbers and their natures are listed in the website of Wolfgang Steinicke, www.klima-luft.de/steinicke/index_e.htm. But from the type objects that Herschel describes in a little more detail, it is clear that *most* of the objects that he is talking about in the later articles (after Article 5) are not gaseous objects at all, but galaxies. There is, however, an occasional bright and compact galactic nebulosity (e.g. the Trifid in Article 11) and of course Article 30 relates to planetary nebulae. A few other objects in the later articles on "condensed nebulae" are also planetary nebulae.

It might seem strange that Herschel should see so many galaxies as essentially star-like objects or "very much brighter towards the middle", with a faint halo or "chevelure" about the bright nucleus. It must be remembered that Herschel was seeing only the brighter central regions of galaxies, not the faint outer parts and spiral arms recorded on modern photographs. Now the familiar textbook reproductions of galaxies are very misleading in suggesting a more-or-less uniformly bright central area from which (in spirals) the arms emerge. On visual inspection, or on good original negatives of short exposure, many galaxies (perhaps most) show a rapid increase in brightness to an astonishingly small and bright central nucleus of almost stellar dimensions. This feature is almost always lost in photographic reproduction, and is one with which many modern astronomers who have not worked on galaxies are unfamiliar.

In this paper of 1811 Herschel is further from the truth (in that he supposes galaxies to be of a gaseous nature and condensing) than he was in 1789, when he thought that most small nebulous objects were made of unresolved stars. But although his interpretation is incorrect, the observations are perfectly sound, and Herschel may fairly be said to have been the first person to discover (as we should now say) that many galaxies have a bright star-like nucleus.

The extraordinary nature of a few of these nuclei was first recognized in 1943 by Carl Seyfert (1911–60) and the type came to be known as 'Seyfert galaxies'. From the 1950s, with the growth of radio astronomy and the availability of larger optical telescopes, the nuclei of galaxies in general were increasingly studied and this led to the concept of 'high-energy astrophysics' and the discovery of objects such as quasars, pulsars, and black holes, which now constitute a large part of astrophysics in the present century. – D.]

8th classes of my Catalogues, and many of the *Connoissance des Temps*, have as yet been left unnoticed. Several other objects, in which stars and nebulosity are mixed, such as nebulous stars, nebulae containing stars, or suspected clusters of stars which yet may be nebulae, have not been introduced, as they appeared to belong to the sidereal part of the construction of the heavens, into a critical examination of which it was not my intention to enter in this Paper.

William Herschel.

Slough, near Windsor,
May 26, 1811.

1814: “Observations relating to the sidereal part of the heavens”

Astronomical Observations relating to the sidereal part of the Heavens, and its Connection with the nebulous part: arranged for the purpose of a critical Examination[1]

Read February 24, 1814.

In my paper of observations of the nebulous part of the heavens, I have endeavoured to shew the probability of a very gradual conversion of the nebulous matter into the sidereal appearance. The observations contained in this paper are intended to display the sidereal part of the heavens, and also to shew the intimate connection between the two opposite extremes, one of which is the immensity of the widely diffused and seemingly chaotic nebulous matter; and the other, the highly complicated and most artificially constructed globular clusters of compressed stars.

The proof of an intimate connection between these extremes will greatly support the probability of the conversion of the one into the other; and in order to make this connection gradually visible, I have arranged my observations into a series of collections, such as I suppose will best answer the end of a critical examination.

[1] [*Phil. Trans.*, **104** (1814), 248–284.]

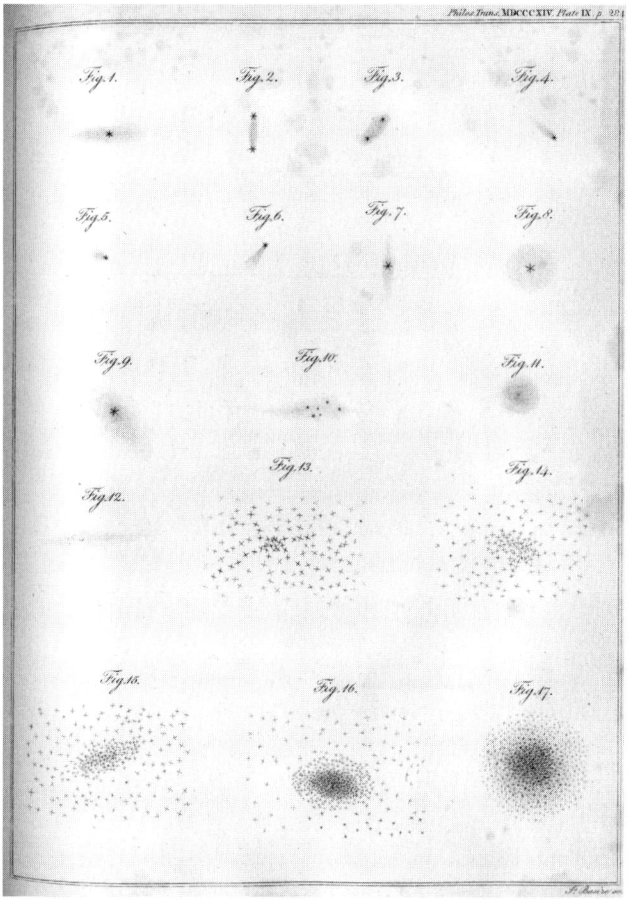

1. Of Stars in remarkable situations with regard to Nebulae[2]

2. Of two Stars with nebulosity between them[3]

A more remarkable situation than the former is that of two stars with nebulosity between them, or both included in the same nebulosity.

III. 67[4] is "An extremely faint nebulosity extended from one star to a smaller one, at the distance of about 2 minutes south of the former." See fig. 2.

[2] [These are galaxies close to foreground stars. – D.]

[3] See nineteen double stars joined by intermediate nebulosity II, 16. 706. 732. III, 19. 32. 67. 68. 113. 126. 182. 200. 312. 376. 540. 637. 757. 785. 820. 854. [It is surprising that Herschel does not include IV. 30 = NGC 4861 (an edge-on galaxy), although he describes it as "two stars dist. 3′ connected with a very faint narrow nebulosity". – S.]

[4] [H. III. 67 = NGC 3473, S galaxy between two foreground stars. – D.]

II. 706.[5] "Two considerable stars are involved in a very faint nebulosity of 3 or 4 minutes in extent." See fig. 3.

Here I have referred to 19 instances, where two stars have an extended nebulosity between them, or at least are both contained within it. Now, if we were to enter into a calculation of chances to investigate the probability that in every one of these 19 objects, the stars and the nebulosity should be unconnected, we should have to consider that in order to produce this appearance by three objects at a distance from each other, it would be required that every one of them should be precisely in a given line of sight, and that the nebulosity should not only be in the middle of them, but that it also should be extended from the situation of one star to that of the other; and that all this should happen in the confined space of a few minutes of a degree; which cannot be probable. Then, if on the other hand we recollect that in the 8th, 9th, and 10th articles of my paper on the nebulous part of the heavens, I have given 139 double nebulae joined by nebulosity between them, and that we have now before us 19 similar objects, with no other difference than that instead of nebulae we have stars with nebulosity remaining between them, should we not surmise that possibly these stars had formerly been highly condensed nebulae, like those that have been mentioned, and were now by gradually increasing condensation turned into small stars; and may not the nebulosity still remaining shew their nebulous origin?

When to this is added that we also have an account of 700 double stars entirely free from nebulosity,[6] many of which are probably at no great real distance from each other, it seems as if we had these double objects in three different successive conditions: first as nebulae; next as stars with remaining nebulosity; and lastly as stars completely free from nebulous appearance.

3. Of Stars with nebulosities of various shapes attached to them

When a nebula seems to be joined to a star, or closely pointing to it, the manner of its appearance deserves our attention. Here follow three different sorts of such conjunctions.[7]

First sort; I. 143[8] "On the north preceding side of a pretty bright star is a considerable, bright nebulosity. It is joined to the star so as to appear like a brush to it." See fig. 4.

[5] [H. II. 706 = NGC 7538, a small compact gaseous (galactic) nebulosity with two stars involved. – D.]

[6] See Phil. Trans. for 1782, page 112; and for 1785, page 40.

[7] See fourteen stars connected with nebulae. Nine with a brush I, 143. II, 214. 683. III, 643. IV, 10. 17. 29. 40. 77. Two with a puff IV, 3. 4. Three with fan-shaped nebulosity IV, 2. 35. 66.

[8] [H. I. 143 = NGC 4900, Sc galaxy near star. – D.]

Second sort; IV. 4[9] "A very small star has an extremely faint, and very small nebula attached to it in the shape of a puff." See fig. 5.

Third sort; IV. 35[10] "A small star has a small, faint, fan-shaped nebulosity joining to it on the north preceding side." See fig. 6.

Here we have a list of fourteen objects,[11] in which the probability of a union between the nebulosities and the stars will gradually become more apparent. With regard to the first nine, the particularity of their construction is already very pointed: the conditions are that the nebulosity must be extended; the direction of its extension must be exactly towards the star, and it must also be apparently just near enough to touch it; but that all this should happen cannot be probable; whereas a real contact of the objects, held together by mutual gravitation, will readily account for the whole appearance.

In the two next objects there is already some indication of a union between the nebula and the star, for the roundness of the nebulosity appeared to be a little drawn out of its figure towards the star.

But the last three instances, in which the whole mass of nebulous matter is pointedly directed to the stars, and in contact with them, can hardly leave any room for doubting a union between them.

Now if we admit a contact, or union between these nebulae and the stars, it deserves to be remarked that stars, in the situation of these fourteen, cannot have been formed from their adjoining nebulosities; for a gradual condensation of the nebulous matter would have been central; whereas the stars are at the extremity of the nebulae. It is therefore reasonable to suppose that their conjunction must be owing to some motion either of the stars or of the nebulous matter: a mutual attraction might draw them together. In either of these cases it would follow, that if the nebulosity should subside into the star, as seems to be indicated by the assumed form of the fan-shaped nebulae, the star would receive an increase of matter proportional to the magnitude and density of the nebulosity in contact with it. This would give us the idea of what might be called the *growth* of stars.

4. Of Stars with nebulous branches

That an intimate connection between the nebulous matter and a star is not incompatible with their nature will clearly appear by the following

[9] [H. IV. 4 = NGC 3662, galaxy and foreground star. – D.]

[10] [H. IV. 35 = NGC 2610, planetary nebula and nearby star. – D.]

[11] [The fourteen examples listed are of objects of various physical natures. Most are galaxies with fortuitous foreground stars nearby, but a few are stars in the vicinity of gaseous or reflection nebulosity. One is, in the modern sense, "a star with nebulosity attached": H. IV. 2 = NGC 2261, Hubble's variable nebula with the variable star R Monocerotis at its tip. – D.]

instances, in which a union is manifested that cannot be mistaken for a deceptive appearance.[12]

IV, 42[13] is "A star of about the 8th or 9th magnitude with very faint nebulous branches extended in the direction of the meridian: each branch is about one minute in length. Other stars of the same size, and at the same time in view, are free from these branches." See fig. 7.

The three objects to which I have referred shew sufficiently that stars and nebulae may be connected; for a little swelling and increase of light of the branches, at their junction with the star, which generally takes place, seems evidently to be an effect arising from the gravitation of the nebulous matter towards a center, in which the star is situated.

Here again the visible effect of gravitation supports the idea of the *growth* of stars by the gradual access of nebulous matter; for in the present case I may refer to the observations already published in the Phil. Trans. for 1811, where, page 301[14] we have an account of twenty-four extended nebulae, gradually a little brighter in the middle; page 303 there are fifty extended nebulae, with an increased brightness towards the middle;[15] page 304, we have fifty-four extended nebulae, with a much greater accumulation of brightness; page 307 there are seven extended nebulae, in which the central increase of brightness approaches towards the formation of a nucleus. Page 309, we have twenty-seven extended nebulae, in which the central nucleus is already formed; and finally, page 311 contains the account of twenty-three extended nebulae, where the nebulosity seems to have so far subsided into the nucleus, as to leave only two opposite faint branches.[16] Who then that has followed up the gradual condensation of an extended nebula till it appeared in the shape of a bright nucleus with faint branches, and finds now in the center of two such opposite faint branches, instead of a condensed nucleus, a star – who, I may ask, would not rather admit that the nucleus had gradually cleared up in brightness, and assumed the lustre of a star, than have recourse to the most improbable of all hypotheses, that a fortuitous central meeting of a star and a nebula should be the cause of such a singular appearance?

[12] See three stars with nebulous branches IV, 42. 43. 48. [IV. 48 = NGC 3104 in Leo Minor is a large irregular galaxy with a superimposed star. – S.]

[13] [H. IV. 42 = NGC 676, star and E6 galaxy. – D.]

[14] [Above, p. 179.]

[15] [Above, p. 180.]

[16] [Above, p. 181.]

5. Of nebulous Stars[17]

6. Of Stars connected with extensive windings of nebulosity[18]

7. Of small patches consisting of Stars mixed with nebulosity

When a small patch of stars is mixed with nebulosity, there is a possibility of its being a deception arising from their being accidentally in the same line of sight; but it has already been shewn that in such appearances the probability is much in favour of a real union; especially when the objects are numerous;[19] and in that case there are but two ways of accounting for it.

First, admitting from what has been said, that stars may be formed of nebulous matter, it may happen that the nebulosity still mixed with them is some remaining unsubsided part of that from which they were formed; and in the next place, a union of stars and nebulosity, originally at a distance from each other, may have been effected by the motion of either the stars or the nebulosity.

That such motions may happen has been shewn in the third article, which contains instances of the conjunction of stars with nebulosities of which they cannot have been formed, and which must, consequently, have been united by motion. We also know that nebulae are subject to great changes in their appearance, which proves that some of the nebulous matter in their composition must be in motion; instances of which have been given in the luminous nebulosity of the constellation of Orion.[20] It may therefore be easily conceived that any moving patch of nebulous matter must be arrested on its meeting with stars; especially if several of them should happen to be pretty near each other; in which case there will be, as it were, a net spread out for intercepting every nebulosity that comes within the reach of their attraction.

II 304.[21] “Three or four stars of various sizes are mixed with pretty strong nebulosity.”

III 165.[22] “Five or six stars forming a parallelogram, are mixed with very faint milky nebulosity.”

[17] See thirteen nebulous stars IV, 19. 25. 36. 38. 44. 45. 52. 57. 58. 65. 69. 71. 74. [These objects are mostly planetary nebulae, with a few stars involved in diffuse galactic nebulosity. – D]

[18] See three stars connected with diffusions of nebulosity IV, 24. 33. V, 27. [Stars involved in diffuse galactic nebulosity, including H. V. 27 = NGC 2264, the star cluster and complex of gaseous and obscuring matter around 15 Monocerotis. – D.]

[19] See thirty-seven small patches, consisting of stars mixed with nebulosity. [Catalogue numbers not reproduced.]

[20] Phil. Trans for 1811, p. 320.

[21] [H. II. 304 = NGC 2316, star in small patch of reflection nebulosity. – D.]

[22] [H. III. 165 = NGC 7186, Herschel’s object is not identifiable; the area he gives has numerous faint stars around, but no nebulosity. – D.]

III. 697.[23] "Several small stars are contained in faint nebulosity about 3 or 4 minutes long and ¾ broad." See fig. 10.

IV. 75.[24] "Three stars of about the 9th or 10th magnitude are involved in pretty strong milky nebulosity."

This collection of thirty-seven objects, consisting of 2, 3, 4, 5, 6, or more small stars that are mixed with nebulosity, contains a variety of instances in which the effect that has been mentioned of the interception of the nebulous matter may have taken place. It is very obvious that nothing positive can be said about the formation of so many starry-nebulous patches; for unless by long continued observation of the same patches we could be acquainted with every change that may happen in the nebulosity or in the magnitude of the stars which apparently compose them, their real union and construction must remain unknown. We can only hint, that every nebulosity which is carried into the region of a small patch of stars will probably be gradually arrested and absorbed by them, and that thus the *growth of stars* may be continued.

8. Of objects of an ambiguous construction

From objects consisting decidedly of stars, but which either have nebulosity mixed with them, or are in such situations as to be seen in the same line with nebulosity, I proceed to give an account of some others, of which my observations have not ascertained into what order we ought to class them.

It has been remarked, on a former occasion, that clusters of stars, when they are at a great distance, may assume a nebulous appearance.[25] This may be experienced by observing a certain celestial object with a telescope of an inferior space-penetrating power, through which it will be seen as a nebula; whereas with an instrument which has a higher degree of this power, its appearance will be a mixture of nebulosity and stars; and if this power of the telescope is of a still higher order, the stars of the same object will then be distinctly perceived: the nebulosity will no longer be seen, and the object will be entitled to be placed into the rank of clusters of stars.

Other objects there are, where a greater space-penetrating power will only increase the brightness of the nebulosity, and at the same time make the tinge of it more uniformly united and of a milky appearance, which will decide it to be purely nebulous.

[23] [H. III. 697 = NGC 4183, an Sc galaxy. – S.]

[24] [H. IV. 75 = NGC 7129, an open cluster with associated reflection nebula. – S.]

[25] Phil. Trans. for 1811, page 270. [Above, p. 170.]

But when an object is of such a construction, or at such a distance from us, that the highest power of penetration, which hitherto has been applied to it, leaves it undetermined whether it belong to the class of nebulae or of stars, it may then be called ambiguous. As there is, however, a considerable difference in the ambiguity of such objects, I have arranged 71 of them into the following four collections.[26]

The first contains seven objects that may be supposed to consist of stars, but where the observations hitherto made, of either their appearance or form, leave it undecided into which class they should be placed.

Connoiss. 31[27] is "A large nucleus with very extensive nebulous branches, but the nucleus is very gradually joined to them. The stars which are scattered over it appear to be behind it, and seem to lose part of their lustre in the passage of their light through the nebulosity; there are not more of them scattered over the nebula than there are over the immediate neighbourhood. I examined it in the meridian with a mirror 24 inches in diameter, and saw it in high perfection; but its nature remains mysterious. Its light, instead of appearing resolvable with this aperture, seemed to be more milky."

The objects in this collection must at present remain ambiguous.

The next contains 26 nebulous objects, of which the figure has been ascertained to be round or nearly round.

II, 101[28] is "A pretty large, round, extremely faint, easily resolvable nebula. I can almost see the stars in it." See fig. 11.

Connoiss. 57[29] is "An oval nebula with an eccentric oval dark space in the middle; there is a strong suspicion of its consisting of stars. The diameter, measured by the large 10 feet telescope, is 1′ 28″,3."

The globular form of the objects in this collection, which is deduced from their round figure, will so far ascertain the manner of their construction, that they must either be still in a condensed state purely nebulous, or else, if

[26] See seventy-one ambiguous objects, in four collections. [Catalogue numbers not reproduced.]

[27] [M31 = NGC 224, the Andromeda galaxy. In 1785 he gave it as an example of a "milky-way": now, although it is "supposed to consist of stars", he regards its nature as "ambiguous". – D.]

[28] [H. II. 101 = NGC 3489, bright galaxy, S0 pec. Note Herschel's use of the word resolvable – "easily resolvable.... I can almost see the stars in it". The stars were far beyond his reach. – D.]

[29] [M57 = NGC 6720, planetary, the ring nebula in Lyra. – D.] [The term 'resolvable', often used by Herschel for what are in fact galaxies, is to be distinguished from 'resolved', often used for globular clusters. 'Resolvable' is more of a guess, when the nebulosity appears mottled. Later visual observers, such as Lord Rosse and E. W. L. Tempel, followed Herschel in this. The effect is often caused by too high a magnification, when the image starts to blur. – S.]

consisting of stars, that they must be already in a far advanced order of compression, and only appear nebulous on account of their very great distance from us. A middle state between the progressive condensation of a globular nebula and a cluster of stars can have no existence; because a globular nebulosity when condensed can only produce a single star. There is, however, a possibility that a mass of nebulous matter in motion may be intercepted by a globular cluster, in which case the nebulosity must soon assume the form of the cluster, and will finally be absorbed by it.

In the third collection I have placed 26 nebulae, which not only are described as easily resolvable, but in most of which some stars have actually been seen.

II, 500[30] is "A very large, easily resolvable, extended, nebulous object. I see a few of the largest stars in it." See fig. 12.

Here the uncertainty in which the descriptions leave us, is that the objects in this collection may be either clusters of stars mixed with nebulosity, or that in consequence of the great distance and compression of the small stars composing a cluster which contains no nebulosity, it may put on the nebulous appearance.

The fourth collection contains 12 nebulous objects, of which the description makes it probable that they belong to the order of clusters of stars.

I, 249[31] is "A considerably bright extended nebula about 4′ long and 2′ broad; it is easily resolvable, and I suppose with a higher power and longer attention the stars would become visible. It is brighter about the middle."

Connoiss. 100[32] is "A nebula of about 10′ in diameter, but there is in the middle of it, a small, bright cluster of supposed stars."

9. Of the sidereal part of the Heavens

The foregoing observations have proved the intimate connection between the nebulous and sidereal condition; and although in passing from one to the other we have met with a number of ambiguous objects, it has been seen that the apparent uncertainty of their construction is only the consequence of the want of an adequate power in our telescopes, to shew them of their real form. We have indeed no reason to expect that an increase of light and distinctness of our telescopes would free us from ambiguous objects; for by improving our power of penetrating into space, and resolving those which we have at present, we should probably reach so many new objects that others,

30 [H. II. 500 = NGC 4535, Sc galaxy. – D.]

31 [H. I. 249 = NGC 2742, Sc galaxy. – D.]

32 [M100 = NGC 4321, bright Sc galaxy. It seems possible that Herschel saw some of the bright condensations that constitute the extreme inner parts of the spiral arms near the nucleus – features lost in normal half-tone reproductions. – D.]

of an equally obscure construction, would obtrude themselves, even in greater number, on account of the increased space of the more distant regions of their situation.

From stars mixed with nebulosity we are now to direct our attention to the purely sidereal part of the heavens; and as stars are the elementary parts of sidereal constructions, it will be proper to review what we know of their nature. Having already entered upon this subject in a former paper at some length,[33] I shall only give a few additional observations, with a summary outline of the former arguments.

The intensity of the light of a star of the first magnitude may be compared with solar light, by considering, that if the sun were removed to the distance at which we generally admit the brightest stars to be from us, its visible diameter could not exceed the 215th part of a second; and its appearance therefore would probably not differ much from the size and brightness of such stars. By reversing this argument we shall be authorised to conclude, from analogy, that stars, were they near enough, would assume the brightness, and some of them perhaps also the size, of the sun; and the consequences that have been drawn from the observations given in my paper on the nature and construction of the sun,[34] may be legitimately applied to the stars; whence it follows that stars, although surrounded by a luminous atmosphere, may be looked upon as so many opaque, habitable, planetary globes; differing from what we know of our own planets, only in their size, and by their intrinsically luminous appearance.

They also, like the planets, shine with differently coloured light. That of Arcturus and Aldebaran for instance, is as different from the light of Sirius and Capella, as that of Mars and Saturn is from the light of Venus and Jupiter. A still greater variety of coloured star-light has already been shewn to exist in many double stars, such as γ Andromedae, β Cygni,[35] and many more.[36] In my sweeps are also recorded the places of 9 deep garnet, 5 bright garnet, and 10 red coloured stars, of various small magnitudes from the 7th to the 12th.

By some experiments, on the light of a few of the stars of the first magnitude,[37] made in 1798, by a prism applied to the eye-glasses of my reflectors, adjustable to any angle and to any direction, I had the following analyses.

[33] Phil. Trans. for 1795, page 68.

[34] [William Herschel, "On the nature and construction of the Sun and fixed stars", *Phil. Trans.*, **85** (1795), 46–72.]

[35] [Albireo, one of the best examples of a double star with different colours. The brighter yellow star contrasts with its fainter blue companion. – S.]

[36] See Catalogue of double stars Phil. Trans. for 1782, III, 5. V, 5, &c.

[37] [Spectroscopy is of course a fundamental tool of modern astrophysics. Prismatic experiments on starlight were suggested to Herschel in 1783 by one Thomas Collinson who spent the night at Datchet. In spite of letters from Collinson (20 June 1783), Watson

The light of Sirius consists of red, orange, yellow, green, blue, purple, and violet.

α Orionis contains the same colours, but the red is more intense, and the orange and yellow are less copious in proportion than they are in Sirius.

Procyon contains all the colours, but proportionally more blue and purple than Sirius.

Arcturus contains more red and orange and less yellow in proportion than Sirius.

Aldebaran contains much orange, and very little yellow.

α Lyrae contains much yellow, green, blue, and purple.[38]

The similarity of the general construction of the sun, the stars, and the planets, is also much supported by the periodical variations of the light of the stars observed in many of them;[39] for these variations can only be satisfactorily accounted for by admitting such stars to have a rotatory motion on their axes, like that which the sun and the planets are known to have.[40]

10. Of the aggregation of Stars

11. Of irregular Clusters

12. Of Clusters variously extended and compressed

13. Of Clusters of Stars of a peculiar description

14. Of differently compressed Clusters of Stars

15. Of the gradual concentration and insulation of Clusters of Stars

16. Of globular Clusters of Stars[41]

17. Of more distant globular Clusters of Stars[42]

(31 August 1783), and Collinson again (5 December 1794), Herschel took little interest in making observations which he had no hope of interpreting.]

38 [Herschel's qualitative descriptions of the spectra are in reasonable agreement with the spectral types: Sirius (A1), α Lyrae (A0), Procyon (F5) showing "all the colours" including violet; Aldebaran (K5), Arcturus (K0), and α Orionis (M2) more light in the red and orange. – D.]

39 See Mr. Pigott's Catalogue of variable stars Phil. Trans. for 1786, page 191.

40 See Remarks on the rotatory motion of stars on their axes, Phil. Trans. for 1796, page 456.

41 [Articles 10–16 are accounts of galactic and globular clusters mostly fully resolved into stars by Herschel. See Figs 14–17. – D.]

42 [Eight objects, comprising five globular clusters, one galactic cluster, one Sb and one Sc galaxy. – D.]

18. Of still more distant globular Clusters of Stars[43]

19. Of a recurrence of the ambiguous limit of observation

In the 16th article I have given a description of the most magnificently constructed sidereal systems; and very little doubt can be entertained but that the objects of the 17th and 18th articles are of the same nature, and are only less beautiful in their appearance as they are gradually more remote. It has already been shewn in the 8th article, that in passing from faint nebulosity to the suspected sidereal condition, we cannot avoid meeting with ambiguous objects, to which I must now add, that the same critical situation will again occur, when from the distinctly sidereal appearance we endeavour to penetrate gradually farther into space. In consequence of this remark, it seems probable that among the numerous globular nebulae which have been given in my last paper, many beautiful clusters of stars may lie concealed. To this we may add, that several of the great number of objects which have been given as stellar nebulae, and are probably at a still greater distance from us, may be the last glimpses we can have of such clusters of stars as the 77th of the Connoissance des Temps,[44] which will nearly put on the stellar appearance when it is viewed in a very good common telescope.

This ambiguity, however, being the necessary consequence of the faintness or distance of objects, when seen through telescopes that are not sufficiently powerful to shew them as they are, will not affect any of the arguments that have been used to establish the existence of a clustering power, the effects of which have gradually been traced from the first indication of clustering stars, through irregular as well as through more artificially arranged clusters, up to the beautiful globular form.

The extended views I have taken, in this and my former papers, of the various parts that enter into the construction of the heavens, have prepared the way for a final investigation of the universal arrangement of all these celestial bodies in space;[45] but as I am still engaged in a series of observations for ascertaining a scale whereby the extent of the universe, as far as it is possible for us to penetrate into space, may be fathomed, I shall conclude this paper by pointing

[43] [Five objects, which Herschel correctly identifies as globular clusters although he had not seen the individual stars. – D.]

[44] [M77 = NGC 1068. Not a cluster of stars, but a Sb galaxy. It has a very bright small nucleus. – D.]

[45] [William Herschel, "Astronomical observations, and experiments tending to investigate the local arrangment of the celestial bodies in space", *Phil. Trans.*, **107** (1817), 302–331, and "Astronomical observations and experiments, selected for the purpose of ascertaining the relative distances of clusters of stars ...", *Phil. Trans.* **108** (1818), 427–490.]

out some inferences which the continuation of the action of the clustering power enables us to draw from the observations that have been given.

20. Of the breaking up of the milky way

The milky way is generally represented in astronomical maps as an irregular zone of brightness encircling the heavens, and my star gages have proved its whitish tinge to arise from accumulated stars, too faint to be distinguished by the eye. The great difficulty of giving a true picture of it is a sufficient excuse for those who have traced it on a globe, or through the different constellations of an *Atlas Coelestis*, as if it were a uniform succession of brightness. It is, however, evident that, if ever it consisted of equally scattered stars, it does so no longer; for, by looking at it in a fine night, we may see its course between the constellations of Sagittarius and Perseus affected by not less than eighteen different shades of glimmering light, resembling the telescopic appearances of large easily resolvable nebulae; but in addition to these general divisions, the observations detailed in the preceding pages of this paper, authorise us to anticipate the breaking up of the milky way, in all its minute parts, as the unavoidable consequence of the clustering power arising out of those preponderating attractions which have been shewn to be everywhere existing in its compass.

One hundred and fifty-seven instances have been given of clusters situated within the extent of the milky way, and their places are referred to in nine preceding articles. They may also be found in BODE's Atlas Coelestis, whose delineation of this bright zone I have taken for a standard. To these must be added 68 more, which are in the less rich parts, or what may be called the vanishing borders of the milky way: for this immense stratum of stars does not break off abruptly, as generally represented in maps, but gradually becomes invisible to the eye when the stars are no longer sufficiently numerous to cause the impression of milkiness.

Now, since the stars of the milky way are permanently exposed to the action of a power whereby they are irresistibly drawn into groups, we may be certain that from mere clustering stars they will be gradually compressed through successive stages of accumulation, more or less resembling the state of some of the 263 objects by which, in the tenth and six succeeding articles, the operation of the clustering power has been laid open to our view, till they come up to what may be called the ripening period of the globular form, and total insulation; from which it is evident that the milky way must be finally broken up, and cease to be a stratum of scattered stars.

We may also draw a very important additional conclusion from the gradual dissolution of the milky way; for the state into which the incessant action of the clustering power has brought it at present, is a kind of chronometer that may be used to measure the time of its past and future existence; and although we do not know the rate of going of this mysterious chronometer, it is nevertheless certain, that since the breaking up of the parts of the milky way affords a proof that it cannot last for ever, it equally bears witness that its past duration cannot be admitted to be infinite.

[WILLIAM HERSCHEL.]

Select bibliography

J. A. Bennett, "Herschel's scientific apprenticeship and the discovery of Uranus", in G. E. Hunt (ed.), *Uranus and the Outer Planets* (Cambridge, 1982), 35–53.

"'On the power of penetrating into space': The telescopes of William Herschel", *JHA*, **7** (1976), 75–108.

J. L. E. Dreyer (ed.), *The Scientific Papers of Sir William Herschel* (2 vols, London, 1912).

Kenneth Glyn Jones, *The Search for the Nebulae* (Chalfont St Giles, UK, 1975).

Mrs John Herschel, *Memoir and Correspondence of Caroline Herschel*, 2nd edn (London, 1879).

Michael Hoskin, *Caroline Herschel's Autobiographies* (Cambridge, 2003).

Discoverers of the Universe: William and Caroline Herschel (Princeton, 2011).

"Herschel's 40ft reflector: Funding and functions", *JHA*, **34** (2003), 1–32.

"Nebulae, star clusters and the Milky Way: From Galileo to Herschel", *JHA*, **39** (2008), 363–396.

Stellar Astronomy: Historical Studies (Chalfont St Giles, UK, 1982).

The Herschel Partnership: As Viewed by Caroline (Cambridge, 2003).

The Herschels of Hanover (Cambridge, 2007).

"Unfinished business: William Herschel's sweeps for nebulae", *History of Science*, **43** (2005), 305–320.

"Vocations in conflict: William Herschel in Bath, 1766–1782", *History of Science*, **41** (2003), 315–333.

Michael Hoskin and David Dewhirst, "William Herschel and the prehistory of stellar spectroscopy", *JHA*, **37** (2006), 393–403.

Constance A. Lubbock, *The Herschel Chronicle: The Life-story of William Herschel and his Sister Caroline* (Cambridge, 1933).

Roy Porter, "William Herschel, Bath, and the Philosophical Society", in G. E. Hunt (ed.), *Uranus and the Outer Planets* (Cambridge, 1982), 23–34.

Simon Schaffer, "The great laboratories of the universe: William Herschel on matter theory and planetary life", *JHA*, **11** (1980), 81–111.

"Uranus and the establishment of Herschel's astronomy", *JHA*, **12** (1981), 11–26.

Wolfgang Steinicke, *Observing and Cataloguing Nebulae and Clusters: From Herschel to Dreyer's New General Catalogue* (Cambridge, 2010).

Index